SpringerBriefs in World Mineral Deposits

Editor-in-chief

Antoni Camprubí, Instituto de Geología, Universidad Nacional Autónoma de México, México, Distrito Federal, México

Series editors

José María González Jiménez, Departamento de Geología, Universidad de Chile, Santiago de Chile, Chile

Francisco Javier González, Recursos Geológicos Marinos, Instituto Geológico y Minero de España, Madrid, Spain

Leo J. Millonig, Würzburg, Germany

John F. Slack, U.S. Geological Survey (Emeritus), Farmington, ME, USA

T0172163

The *SpringerBriefs in World Mineral Deposits* book series seeks to publish monographs or case studies focused on a single mineral deposit or a limited group of deposits (sub-regional level), with regard to their mineralogy, structure, geochemistry, fluid geochemistry, and any other aspect that contributes to explaining their formation. This series is aimed at academic and company researchers, students, and other readers interested in the characteristics and creation of a certain deposit or mineralized area. The series presents peer-reviewed monographs.

The Springer Briefs in World Mineral Deposits series includes both single and multi-authored books. The Series Editors, Prof. Antoni Camprubí (UNAM, Mexico), Dr. Francisco Javier González, Dr. Leo Millonig, Dr. John Slack (SGA publications editor) and Dr. José María González Jiménez are currently accepting proposals and a proposal form can be obtained from our representative at Springer, Dr. Alexis Vizcaino (Alexis.Vizcaino@springer.com).

More information about this series at http://www.springer.com/series/15086

Thomas Dittrich · Thomas Seifert ·
Bernhard Schulz · Steffen Hagemann ·
Axel Gerdes · Jörg Pfänder

Archean Rare-Metal Pegmatites in Zimbabwe and Western Australia

Geology and Metallogeny of Pollucite Mineralisations

 Springer

Thomas Dittrich
Division of Economic Geology
and Petrology, Institute of Mineralogy
TU Bergakademie Freiberg
Freiberg, Sachsen, Germany

Thomas Seifert
Division of Economic Geology
and Petrology, Institute of Mineralogy
TU Bergakademie Freiberg
Freiberg, Sachsen, Germany

Bernhard Schulz
Division of Economic Geology
and Petrology, Institute of Mineralogy
TU Bergakademie Freiberg
Freiberg, Sachsen, Germany

Steffen Hagemann
Centre for Exploration Targeting, School
of Earth Sciences
The University of Western Australia
Crawley, WA, Australia

Axel Gerdes
Department of Geosciences
Goethe University Frankfurt
Frankfurt am Main, Hessen, Germany

Jörg Pfänder
Ar-Ar-Lab/Division of Tectonophysics
Institute for Geology
TU Bergakademie Freiberg
Freiberg, Sachsen, Germany

Additional material to this book can be downloaded from http://extras.springer.com.

ISSN 2509-7857   ISSN 2509-7865   (electronic)
SpringerBriefs in World Mineral Deposits
ISBN 978-3-030-10942-4   ISBN 978-3-030-10943-1   (eBook)
https://doi.org/10.1007/978-3-030-10943-1

Library of Congress Control Number: 2018966846

This Springer imprint is published by the registered company Springer Nature Switzerland AG
The registered company address is: Gewerbestrasse 11, 6330 Cham, Switzerland

Preface

Lithium–cesium–tantalum (LCT) pegmatites are important resources for rare metals, whose demand increased markedly during the past decade. Especially, cesium and its compounds are used in drilling fluid in hydrocarbon exploration, as a catalyst in the chemical industry, and in advanced technology applications as atomic clocks, airbag detonators, high-resolution display screens or as a propellant agent in ion engines. The Cs metal market is actually in the frame of lower tens of thousand kilograms per annum. Data on resources and production are very limited. Cs resources can be categorised into one group of potassium salts, sedimentary rocks and seawater, all with no commercial interest due to low concentrations. The Cs resource group of commercial importance is LCT pegmatites that contain the rare zeolite group cesium mineral pollucite. At present, Cs and reserves of it are known in economic and mineable quantities only from the two LCT pegmatite deposits at Bikita in the Zimbabwe Craton and Tanco in the Superior Craton in Canada, where pollucite occurs in monomineralic mineralisation. Both pegmatite deposits have a comparable regional geological background, as they are hosted within greenstone belts and yielded Neo-Archean ages at ~2625 Ma. A closure of the Tanco Mine for safety reasons and an imminent depletion of the pollucite resources at Bikita stimulated the specialised search for pollucite mineralisation in the frame of the exploration boom for Li and rare metals in pegmatites. In Western Australia, the Meso- to Neo-Archean units in the Yilgarn and Pilbara cratons are known to host many LCT pegmatites, among them world-class economic deposits as Greenbushes (Li-Ta-Sn) and Wodgina (Li-Ta-Sn). Geological mapping programmes by the Geological Survey of Western Australia and the National Geological Survey of Australia, as well as countless prospecting and exploration attempts, could identify a large number of LCT pegmatites. Due to this potential, the Rockwood Lithium GmbH at Frankfurt/Main (Germany) initiates and funds the exploration on LCT pegmatites with a special focus on Cs. The first author participated as investigator in such campaigns between July 2011 and June 2016, where a total of 19 pegmatite districts were inspected and sampled. The second author acted as the principal supervisor of the project. As the search for massive Cs-pollucite mineralisation is influenced by genetic concepts on their formation, the

Bikita pegmatite was included in the study. This book presents the analytical results and their interpretations, generated during the intensive search to find one more deposit of a yet merely unique mineralisation type on the planet Earth.

Freiberg, Germany   Thomas Dittrich
Freiberg, Germany   Thomas Seifert
Freiberg, Germany   Bernhard Schulz
Crawley, Australia   Steffen Hagemann
Frankfurt am Main, Germany   Axel Gerdes
Freiberg, Germany   Jörg Pfänder

Acknowledgements

Many persons have generously given time, advice, encouragement and valuable support for this study. Despite there is a risk for forgetting someone, we will do our best to mention as many as possible and our apologies to those we have missed.

Foremost, we would like to express our tremendous gratitude to Dr. Bettina Munsch, Gerrit Fuelling and Joachim Weinrich of the Chemetall GmbH/Rockwood Lithium GmbH Frankfurt am Main (Germany), for the opportunity and the financial support to perform the present study.

Dr. Gernot Hess is thanked for his supervision and introduction to the Bikita pegmatite field. We are particularly grateful to Bikita Minerals Private Limited, especially Nigel McPhail (†), Andre van der Merwe and Colin Morgan for getting access to the Bikita mining facility and their support during the fieldwork and sample shipment.

Furthermore, we would like to thank Terry Stark, Philip Tornatora (Galaxy Resources Limited) for permissions to enter and sample the mine site at Cattlin Creek. Peter Stockman, Michael Herbert and Glenn Oklay (Global Advanced Metals Pty Ltd) gave permission to work at Wodgina. Also, we acknowledge the members of the Strindberg family for getting access to their deposits and tenements, and support at Mount Deans. The Coolgardie Gem and Mineral Club Inc. allowed working at Londonderry. The permissions to enter the area for the other pegmatite occurrences were granted by the current landowners. Catherine Nykakecho and Celia Guergouz are thanked for the assistance during the fieldwork in Western Australia. Logistics and organisation of the fieldwork 2011 and 2012 in Western Australia were based at the Centre for Exploration Targeting (CET) at the University of Western Australia (UWA) in Perth. The CET contributed with arrangement of permissions from landowners in order to enter farmlands and prospect areas and the organisation of field equipment for remote desert areas. We gratefully acknowledge Dr. Walter Witt (CET) and Dr. Marcus Sweetapple (CSIRO, Perth/Western Australia) for their assistance during the field work and helpful discussions. The Mineral Museum of Western Australia is acknowledged for providing some sample material of the Londonderry pollucite specimen.

A huge debt of gratitude goes to Prof. Dr. Jens Gutzmer (Helmholtz Institute Freiberg for Resource Technology, HIF) for his continued support, interest and encouragement during the course of this study. Andreas Bartzsch (HIF) and his staff and Dr. Michael Magnus (Institute for Geology, TU Bergakademie Freiberg) and his team are kindly thanked for the preparation of numerous samples.

The authors are further indebted to many fellow researchers and scientists who graciously gave us their energy and time during laboratory work or for providing valuable datasets. Dr. Wolfgang Dörr at Institut für Geowissenschaften, Goethe-Universität Frankfurt/Main, delivered TIMS analyses. Dr. Tomáš Magna from the Czech Geological Survey provided lithium isotope data. Dr. David Banks from the School of Earth and Environment, University of Leeds (UK), contributed fluid geochemistry data. Dr. Volker Lüders from the Helmholtz Centre Potsdam GFZ German Research Centre for Geosciences analysed the carbon isotopes in sample gas from fluid inclusions. Sabine Gilbricht and Dr. Dirk Sandmann (Department of Economic Geology and Petrology, TU Bergakademie Freiberg) gave assistance during the work with the scanning electron microscopes and the Mineral Liberation Analysis software package at the Geometallurgy Laboratory Freiberg.

Finally, we would like to thank numerous of students, including Lisa Richter, Matthias E. Bauer, Albert Gruber, Julia Schönfeld, Dominique Brising, Henning Scheibert, Leonie Josten, Julia Schmiedel and Florian Will, for their scientific interest to work on LCT pegmatites, as well as for assistance and logistical support.

Contents

1 Introduction to Archean Rare-Metal Pegmatites 1
  1.1 The Alkali Metal Cesium 1
  1.2 Mineralogy and Geochemistry of Cesium 2
  1.3 Mineralogy of Pollucite $(Cs, Na)_2Al_2Si_4O_{12} \times H_2O$ 4
  1.4 General Characteristics and Classification of Pegmatites 9
  1.5 Controls on Pegmatite Formation and Evolution 13
  1.6 Pegmatite Age Distribution and Continental Crust Formation 15
  References 16

2 Geological Settings of Archean Rare-Metal Pegmatites 23
  2.1 Zimbabwe Craton 23
    2.1.1 Pegmatites in the Zimbabwe Craton 25
    2.1.2 The Bikita LCT Pegmatite Field in the Masvingo
            Greenstone Belt 26
    2.1.3 Deposit Geology of Bikita 28
  2.2 Yilgarn Craton 30
    2.2.1 Terranes in the Yilgarn Craton 33
    2.2.2 Tectono-Magmatic Evolution 34
    2.2.3 Pegmatites in the Yilgarn Craton 38
    2.2.4 Geological Setting of the Londonderry Pegmatite
            Field 39
    2.2.5 Geological Setting of the Mount Deans Pegmatite
            Field 42
    2.2.6 Geological Setting of the Cattlin Creek Pegmatite
            Deposit 43
  2.3 Pilbara Craton 45
    2.3.1 Tectonic Model of Archean Evolution 48
    2.3.2 The Wodgina Pegmatite District in the Pilbara Craton . . . 50
  References 54

3 Petrography and Mineralogy 61
   3.1 Minerals in LCT Pegmatites 62
      3.1.1 Feldspars 62
      3.1.2 Quartz 64
      3.1.3 Mica 64
      3.1.4 Pollucite 65
      3.1.5 Petalite 67
      3.1.6 Spodumene 67
      3.1.7 Beryl 68
      3.1.8 Tourmaline 70
      3.1.9 Apatite 71
      3.1.10 Ta-, Nb- and Sn-Oxides 71
   3.2 Reconstruction of the General Crystallisation Sequence 72
   References 74

4 Geochemistry of LCT Pegmatites 77
   4.1 Major and Selected Trace Elements 77
   4.2 Rare Earth Elements 84
   References 86

5 Geochronology of Archean LCT Pegmatites 87
   5.1 Bikita Pegmatite Field 88
   5.2 Londonderry Pegmatite Field 89
   5.3 Mount Deans Pegmatite Field 92
   5.4 Cattlin Creek Pegmatite 92
   5.5 Wodgina Pegmatite 93
   References 94

6 Radiogenic and Stable Isotopes, Fluid Inclusions 95
   6.1 Whole Rock Sm–Nd Isotope Compositions 95
   6.2 Lithium Isotope Analysis on Selected Mineral Phases 97
   6.3 Fluid Inclusion and Pressure-Temperature Data 98
      6.3.1 Fluid Inclusion Studies 99
   References 101

**7 Genesis of Massive Pollucite Mineralisation in Archean LCT
Pegmatites** 103
   7.1 Pegmatite Ages and Potential Source Granites 103
   7.2 Mineralogical and Geochemical Characteristics of LCT
Pegmatites 105
   7.3 Late Stage Hydrothermal Processes 109
   7.4 Structural Setting of Massive Pollucite Mineralisation
in LCT Pegmatites 110

7.5 Concepts for the Formation of Massive Pollucite
Mineralisation 112
    7.5.1 Melt Immiscibility and Separation.................. 112
    7.5.2 Extreme Cesium Enrichment in Massive Pollucite
          Mineralisation 115
    7.5.3 The Role of Feldspars for Cesium Enrichment......... 116
    7.5.4 Cesium Enrichment by Late Stage Hydrothermal
          Processes.................................... 116
7.6 Genetic Model for the Formation of Massive Pollucite
Mineralisation 118
7.7 Implications to the Exploration for Cs-Bearing Pegmatites 121
References 122

Chapter 1
Introduction to Archean Rare-Metal Pegmatites

1.1 The Alkali Metal Cesium

The element Cesium was first described by the German chemist Robert Wilhelm Bunsen and the physicist Gustav Robert Kirchhoff during the investigation of mineral water from Dürkheim (Kirchhoff and Bunsen 1861). Cesium was the first element that was discovered by emission spectroscopy and is characterised by a set of bright blue lines. Due to this, Kirchhoff and Bunsen (1861) named the newly discovered element *caesius*, the Latin word for "sky-blue". Cesium is a chemical element belonging to the subgroup Ia of the alkali-metals in the periodic table. It is a silvery-gold, soft, extremely reactive and pyrophoric metal. Cs has physical and chemical properties similar to other alkali metals like Rb or K, with a large ionic radius of 1.65 Å, and belongs to the large ion lithophile elements (LILE). It has rather low melting (28.7 °C) and boiling points (668 °C), like Hg (Bick et al. 2010). Cesium predominantly forms compounds with halogens (CsF, CsCl) and oxygen as Cs_2O (Wedepohl 1978). Natural Cs compounds as CsCl are only faintly toxic, and are not considered as a significant environmental hazard (Pinsky et al. 1981). As Cs exhibits an extremely low ionisation potential, it is far more reactive than Li, Na or K and still pronouncedly more reactive than Rb. When exposed to air, an explosion-like oxidation reaction will form cesium superoxide CsO_2. In contact with water, Cs reacts vigorously and forms cesium hydroxide and hydrogen gas, with the latter igniting spontaneously (Bick et al. 2010). Although Cs has a total of 39 known isotopes, with mass numbers ranging from 112 to 155, only the ^{133}Cs is natural (Audi et al. 2003).

The cesium market is very small. As a result, data on Cs resources and production are not available or very limited. According to USGS-Cs-2017 the main pollucite zone at Tanco LCT pegmatite deposit in Canada comprises ~120,000 tons of Cs_2O contained in pollucite ore, at ore grades of ~23.3 wt% Cs_2O. Additional reserves are reported from Zimbabwe (~60,000 t) and Namibia (30,000 t). The annual world consumption in 1978 was about 20 t of Cs, as metal and in compounds, and increased during the last decades. However, the market for Cs metal is still in the lower tens of thousands kilograms range per annum.

The first comprehensive studies on the distribution and behaviour of Cs on the Earth concerned its distribution in minerals and rocks (Horstman 1957; Wedepohl 1978; Barnes et al. 2012), or its interactions in host rocks of natural hydrothermal systems (Ellis and Mahon 1977; Keith et al. 1983). More recent studies concerned the behaviour of Cs during metamorphism and melting in subduction zone settings (Hart and Reid 1991; Hall et al. 1993; Bebout et al. 2007; Xiao et al. 2012), or in dependency of the water content of granitic melts (Watson 1979). The behaviour of Cs during the magmatic to hydrothermal processes, and the concepts of LCT pegmatite formation were later picked up and expanded in numerous studies by several working groups (Černý et al. 1985; Icenhower and London 1995; London 2008; Thomas and Davidson 2012).

According to McDonough et al. (1992), about 55% of Cs on Earth occurs in the continental crust, 4% of the Cs is incorporated in the residual mantle, and the remaining 40% of the element remain in the less depleted mantle reservoir. The concentration of Cs in the primitive mantle was estimated by Lyubetskaya and Korenaga (2007) to be 16 ppb. The average concentration of Cs in the upper crust has been estimated to be 4–5 ppm (Taylor and McLennan 1985; Rudnick and Gao 2014), while the lower crust has a Cs abundance of 0.5 ppm Cs (McDonough et al. 1992). Based on a bulk composition derived from 70% lower crust and 30% upper crust, an average content of about 2.1 ppm Cs can be estimated for the total continental crust, and compared to estimated 0.023 ppm Cs for the silicate earth. Ultramafic rocks contain <1 ppm Cs, however, some Archean mantle eclogites and peridotites have remarkably higher Cs of up to 9 ppm (McDonough et al. 1992). The average abundance in igneous rocks ranges from <1 ppm in mafites to about 10 ppm Cs in granitoids (Hall et al. 1993). Some highly fractionated evolved Permo-Carboniferous granites and lamprophyres of the Erzgebirge (Germany) can contain up to 204 ppm and 104 ppm Cs, respectively (Seifert 2008). Lower Permian Sn-F-enriched rhyolitic ignimbrites of the Sub-Erzgebirge basin show Cs contents up to 174 ppm (Seifert 2008). In Zinnwaldite from the Li-Sn greisen deposit Zinnwald (Germany) Cs contents of up to 750 ppm were measured (Neßler et al. 2017). The high-F topaz rhyolite from the Tertiary Spor Mountain Formation (Utah, USA) shows a Cs enrichment of up 87 ppm (Dailey et al. 2018). Sedimentary rocks have an average between 4 and 12 ppm Cs, and clay minerals like kaolinite, bentonite, illite show a modest enrichment up to 17 ppm Cs (Horstman 1957). Oceanic water contains only 0.37 ppb Cs, but in potassic salt an average of 56 ppm Cs can be found (Osichkina 2006).

1.2   Mineralogy and Geochemistry of Cesium

As the Cs shares many properties with other alkali metals as Na, K or Rb, it occurs as trace element in feldspar or mica. Certain geological processes are capable to enrich

Cs to several thousand ppm so that specific conditions can lead to the formation of discrete Cs minerals, which however are very rare. Among the major elements, only the K can be substituted by Cs (Černý et al. 1985). As Cs is almost incompatible during magmatic crystallisation, it becomes enriched in the residual melts. In granitic systems, the Cs content in coexisting phases decreases in the sequence biotite—muscovite—K-feldspar. The relative scarcity of mica in granites shifts the whole rock concentration of Cs into K-feldspar (Černý et al. 1985; Hall et al. 1993). In consequence, K-feldspar, muscovite and lepidolite that crystallise possibly from highly fractionated and geochemical specialised granitic melts enriched in Cs, Rb, Li, as in LCT pegmatite systems, can accommodate much higher amounts of Cs. Only in the most complex and pollucite bearing LCT pegmatites, K-feldspar can contain up to 0.29 wt% Cs_2O. The amount of Cs that substitutes into albite is very low and can reach up to 0.11 wt% Cs_2O in LCT pegmatites (Icenhower and London 1995). In pollucite-bearing LCT pegmatites, muscovite has up to 0.2 wt%, and Li-muscovite and lepidolite up to 1.9 wt% Cs_2O (Tindle et al. 2005). Beryl with 1–4 wt% Cs_2O is extremely rare and only known from LCT pegmatites (Černý 1975; Černý and Simpson 1977). Pyroxene and olivine can incorporate up to 0.88 ppm and 3.47 ppm Cs, respectively (Roselieb and Jambon 1997).

Actually 31 Cs minerals (Table A1) are known and approved by the International Mineralogical Association (IMA). Most of them crystallise in granitic pegmatites or in alkaline complexes at late stage magmatic to hydrothermal processes. Only the zeolite Cs mineral pollucite as part of the analcime-pollucite series, is known to occur in larger and also economic quantities. The other 30 Cs bearing minerals are found in small crystals in interstitial positions and cavities, or are intergrown with other minerals. Nanpingite, the Cs-bearing analogue of muscovite, was discovered within pegmatite in the Nanping area (Fujian province, China) and forms distinct platy crystals intergrown with quartz, apatite and montebrasite (Jambor and Vanko 1990). The Cs-bearing analogue of phlogopite, sokolovaite, was first described from pegmatites of the Red Cross Lake area in Canada and then reported from various LCT pegmatite locations worldwide where it occurs with quartz, lepidolite and pollucite (Černý et al. 2003; Wang et al. 2007; Potter et al. 2009).

During chemical weathering and hydrothermal processes, regardless of the rock type, all released Cs should preferentially be absorbed by most clay minerals. Cs sorption is favoured when Na is the dominant competing ion (Wahlberg and Fishman 1962; Merefield et al. 1981). Alkali metals are leached by hydrothermal fluids at temperatures ranging from 100 to 600 °C from the surrounding host rock. Within the Yellowstone hydrothermal field the rhyolitic host rocks contain 2.5–7.6 ppm Cs. In the geothermal fields of New Zealand with ignimbrites, andesites and basalts as well as greywackes, the host rocks contain less than 2 ppm Cs. The reported Cs contents of the hydrothermal waters from the two geothermal areas are comparable with 0.02–2.6 ppm Cs. It was demonstrated that the leachability of the alkali elements depends on temperature, with more Cs leached at elevated temperatures between 400 and 600 °C (Ellis and Mahon 1977; Keith et al. 1983). At the Yellowstone geothermal field, the dissolved Cs in the hydrothermal solutions apparently interacts with analcime which is considerably enriched with up to 3000 ppm Cs. In zones with no analcime, much of the Cs remains in solution, and is later adsorbed on clay minerals. Although it was demonstrated from these hydrothermal areas that Cs can be effectively leached from the host rocks, there remains the potential that a part of the Cs comes from the roof of an underlying silicic magma chamber which triggered the circulation of late stage magmatic fluids (Keith et al. 1983).

In subduction zones two processes are available that can transform Cs from the oceanic crust into the mantle wedge. During metamorphism of subducting sedimentary rocks with 4–12 ppm Cs, the Cs is retained up to upper greenschist facies conditions. In contrast, high grade metamorphic rocks typically are characterised by a pronounced Cs depletion that apparently is decoupled from the behaviour of K and Rb (Hart and Reid 1991). During prograde metamorphism of sedimentary rocks, Cs is incorporated into muscovite and biotite or in phengite. These micas will conserve Cs until their breakdown at depths >40 km. In cold subduction zones, much of the sedimentary LILE is apparently retained to depths of ~60–90 km (Bebout et al. 2007). When hosted in phengite, Cs can be retained even to depths of 300 km at conditions of 95–110 kbar, 750–1050 °C (Xiao et al. 2012). Thus, micas can be an important carrier of Cs, K, Rb and H_2O into the upper mantle at the base of mantle wedges (Melzer and Wunder 2001). As Cs is portioned into mica compared to feldspar, water absent melting will produce a more Cs enriched melt (Hall et al. 1993).

The large ionic radius of Cs (1.65 Å) compared to the other alkali elements governs that it is partitioned into the remaining melt during fractional crystallisation. In consequence, Cs contents of all igneous rocks remain low (1–10 ppm). Modest enrichment of Cs is only reported for leucogranites and pegmatitic granites or fertile granites parental to LCT pegmatites where the Cs can range up to about 200 ppm (Selway et al. 2005). A further significant increase in Cs content is reported from LCT pegmatites that can contain several thousand ppm of Cs and finally can host economic quantities of massive pollucite mineralisation with Cs contents up to 25 wt% Cs_2O (Stilling et al. 2006).

1.3   Mineralogy of Pollucite $(Cs, Na)_2Al_2Si_4O_{12} \times H_2O$

The Cs mineral pollucite was first discovered on the island of Elba by Breithaupt (1846) and analysed by Plattner (1846). Due to its close association to petalite it was named after the Greek myth companion figures Castor and Pollux. Pollucite is classified as tectosilicate and belongs to the zeolite group. It has a general composition of $(Cs, Na)_2Al_2Si_4O_{12} \times H_2O$ and is isostructural to analcime $NaAlSi_2O_6 \times H_2O$ (Barrer and McCallum 1951). It crystallises in cubic, dodecaedrical or trapezohedral crystals with colours ranging from colourless, white or gray and rarely purple, pink or blue. More commonly, pollucite is developed as glassy, colourless to white polycrystalline masses. The analcime structure was described by Taylor (1930) and later redefined for pollucite (Naray-Szabo 1938). The analcime structure consists of an open framework of SiO_4 and AlO_4 tetrahedra with Na and H_2O occupying the large voids in the framework. Pollucite and analcime form a solid

Table 1.1 Compositional ranges observed for the analcime-pollucite solid solution series as obtained from the MinIdent-Win4 database

Mineral	Formula	Range	Density g/cm^3	Cs$_2$O wt%
Analcime	(NaAl)(Si$_2$O$_6$) H$_2$O	Pol$_0$–Pol$_5$	2.21–2.25	0.0–2.23
Cs-Analcime	(CsNaAl)(Si$_2$O$_6$) H$_2$O	Pol$_5$–Pol$_{50}$	2.25–2.58	2.23–22.6
Na-Pollucite	(CsNa)$_2$(Al$_2$Si$_4$O$_{12}$) H$_2$O	Pol$_{50}$–Pol$_{95}$	2.58–2.90	22.6–42.9
Pollucite	(CsNa)$_2$(Al$_2$Si$_4$O$_{12}$) H$_2$O	Pol$_{95}$–Pol$_{100}$	2.90–2.94	42.9–45.2

solution series and are connected via a substitution of Cs$^+$ for Na$^+$ and H$_2$O. The analcime-pollucite series is composed by the endmember pollucite (Pol$_{95}$–Pol$_{100}$); sodian (Na–) pollucite (Pol$_{50}$–Pol$_{95}$); cesian (Cs–) analcime (Pol$_5$–Pol$_{50}$), and the endmember analcime (Pol$_0$–Pol$_5$; Table 1.1). Several miscibility gaps at Pol$_0$–Pol$_9$, Pol$_{55}$–Pol$_{62}$, and Pol$_{82}$–Pol$_{100}$ compositions between the two endmembers, and an end-to-end miscibility between the two endmembers pollucite and analcime have been suggested (Černý 1974). Beger (1969) defined pollucite by tetrahedral Al and Si that form a three-dimensional framework pore system, with multiple cation sites in the cages and pores. The Na is situated between four O atoms of the Si–Al tetrahedron and the two water molecules. The larger Cs atom will remain in the water position. Other structural models for pollucite are provided by Newnham (1967), Yanase et al. (1997), Kamiya et al. (2008) and Gatta et al. (2009). The crystal structure of pollucite shows a rapid thermal expansion between 25 and 400 °C (Kobayashi et al. 1997).

Primary pollucite can crystallise at near solidus temperatures in Cs and F enriched granitic melts (Teertstra et al. 1992). Also pollucite is formed in low temperature "Alpine vein" assemblages in cavities and is interpreted to be formed by leaching and reprecipitation in primary pollucite-bearing LCT pegmatites (Smeds and Černý 1989). The crystallisation conditions of pollucite were experimentally determined to be in the range from 600 to 300 °C (Teertstra et al. 1992). However, the temperature of precipitation in nature from B, F, Li, H$_2$O rich pegmatite forming melts is probably somewhat below this range (London 1990). The chemistry and stability of pollucite-analcime solid solution in the haplogranite system was further experimentally determined between 450 and 850 °C at 200 MPa H$_2$O (London et al. 1998). The addition of Cs via the dissolution of pollucite lowers the haplogranite solidus by 40 °C (to 640 °C) and displaces the minimum melt composition slightly towards SiO$_2$. The Cs content of the melt has to be about 5 wt% Cs$_2$O near the solidus temperature at 640 °C in order to be saturated for crystallising pollucite-analcime solid solution. Nevertheless, London et al. (1998) mentioned that conditions for the direct crystallisation of pollucite (Pol$_{50}$) at 997 °C or even endmember pollucite (Pol$_{100}$) at 2229 °C from a melt are not realised in granitic and LCT pegmatitic melts.

London (1995) suggested that such anomalously Cs enriched melts can be generated by low temperature anatexis of muscovite rich protoliths as mica schists. This will only produce a small fraction of melt at incipient hydrous anatexis and is followed by the breakdown of muscovite to alkali feldspar and corundum or aluminosilicate at higher temperatures. According to London et al. (1998) this process will release

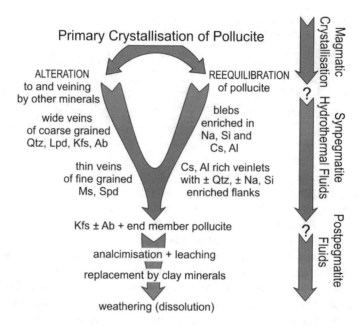

Fig. 1.1 Relative timing of post-crystallisation processes affecting pollucite after a magmatic crystallisation, adopted from Teertstra and Černý (1995); mineral abbreviations after Kretz (1983)

most of the Cs contained in the white mica into a small volume of melt. In contrast, if biotite or cordierite is present in the protolith, a large fraction of Cs is retained in the residual minerals and is released out to the melt over a considerable larger range of temperature (Bea et al. 1994; Icenhower and London 1995). Even though, the average Cs content of mica schists (4–14 ppm Cs; Horstman 1957) is generally low, a further enrichment by the factor of 100 is necessary to accumulate sufficient amount of Cs to crystallise pollucite-analcime solid solution within the model of London et al. (1998). The Cs_2O content within the solid solution series can range between 2.21 and 45.16 wt% Cs_2O (Table 1.1). Sodium poor pollucite close to the ideal endmember composition is only reported on a microscopic scale from a few localities in Europe, Africa and North America (Teertstra and Černý 1995). The H_2O content decreases with decreasing Na content and increasing Cs content (Barrer and McCallum 1951) and is explained by isostructural exchange of Na + H_2O and Cs. The Cs ion is too large to migrate from one interstitial site in the crystal to the next. Consequently, this framework exhibits an ion sieve effect on the Cs ion.

Primary homogenous pollucite (i) crystallises from a residual melt at near solidus conditions of near-equilibrium, or slightly below solidus temperature, at disequilibrium (Fig. 1.1). During subsequent cooling, the homogenous pollucite will exsolve (ii) into Cs, Al and Na, Si enriched components. This step is followed by several generations of veining of progressively Cs–Al enriched veinlets (iii) accompanied by distinct Na–Si enriched outward fading flanks. The next period of pollucite for-

mation is characterised by the formation of coarse veins (iv) that are filled by late
stage crystallisation assemblages, as lepidolite, quartz and albite. This is followed by
various generations of Cs or Na metasomatism (v) that resulted in almost endmember
pollucite or analcime (Teertstra et al. 1992, 1995, 1996). Černý and Simpson (1978)
and Teertstra et al. (1993) reported replacement of pollucite by adularia (vi), analcime (vii) or clay minerals (viii) during subsequent hydrothermal activity (Fig. 1.1).
Wood and Williams-Jones (1993) stated that alteration of pollucite to albite or K-
feldspar upon cooling is only likely to occur if fluids with very high contents of Na
and/or K entered the pegmatites. While stages (i) and (ii) are of magmatic origin,
stages (iii) to (v) are interpreted by Teertstra et al. (1996) to result from syngenetic
hydrothermal fluids. Stages (vi) to (viii) then might result from the vanishing syngenetic hydrothermal fluids. Furthermore, especially stage (viii) could also result
from the activity of post-pegmatite hydrothermal fluids (Fig. 1.1). The process of
dissolution and reprecipitation (steps iii to iv) of pollucite occurs under open-system
hydrothermal conditions at 330 °C, 1.5 kbar and silica-oversaturation (Lagache et al.
1995). When pollucite is dissolved on a local scale and the fluid inherits its Na/(Na +
Cs) ratio, then any reprecipitation from this fluid would generate a pollucite distinctly
enriched in Cs relative to the precursor phase. Teertstra and Černý (1995) stated that
the late interstitial fluid that dissolves the primary pollucite has necessarily a Na
+ Cs content of its own, which will further modify the composition of the solute
and the resulting solid phase. Another important factor of these subsolidus reactions
is the silica-undersaturated chemistry of their local environment. The availability
of liberate quartz in such a system is limited, therefore Teertstra and Černý (1995)
suggested that silica oversaturation is probably controlled by pollucite itself through
Si–Al exchange. They stated further that these reactions cannot be directly compared
to the pollucite + albite equilibrium experiments of Sebastian and Lagache (1990)
in the presence of excess SiO$_2$ and at 750–450 °C and 1.5 kbar. This raises the ques-
tion, if pegmatite melts can be sufficiently enriched by Cs to allow crystallisation
of pollucite, or, more focussed, if crystal fractionation processes alone allow such
a Cs enrichment. Apart from such discrepancies, these studies revealed that pollucite is formed over a large temperature range starting from about 750 °C down to
150 °C, either by direct crystallisation from a granitic melt (Sebastian and Lagache
1990; London et al. 1998), or via precipitation from a hydrothermal fluid (Redkin
and Hemley 2000; Endo et al. 2013; Yokomori et al. 2014; Jing et al. 2016, 2017).
Despite this large temperature range, it is apparently accepted that pollucite is one
of the last minerals formed during the course of the crystallisation of the pegmatite.
An endmember pollucite (Pol$_{100}$) represents an enrichment factor of 213,000 times
of the average 2 ppm Cs content in the continental crust.

Pollucite occurrences are reported from approximately 140 locations worldwide.
However, at present only two LCT pegmatite deposits are known that host economic
quantities of massive pollucite mineralisation. These are the Bikita LCT pegmatite
deposit in Zimbabwe and the Tanco LCT pegmatite deposit in Canada (Fig. 1.2). Both
represent highly evolved LCT pegmatite systems that have a similar Neo-Archean age
between 2650 and 2600 Ma, and are hosted in greenstone belt rocks. The massive
pollucite mineralisation occur in lensoid bodies within the upper portions of the

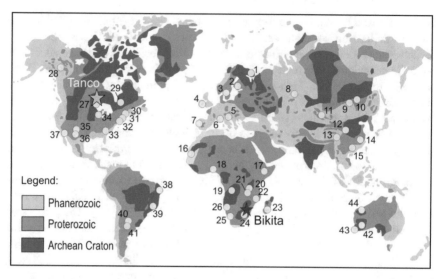

Fig. 1.2 Age of the continental crust and locations of major LCT pegmatite fields and districts (modified after Bradley et al. 2017). Occurrences of massive pollucite (Cs) mineralisations in pegmatites at Tanco and Bikita are marked by stars. **Europe**: 1 Vasin Myl'k (Russia); 2 Varuträsk (Finland); 3 Borkenas-Orust (Sweden); 4 Leinster (Ireland); 5 Koralpe (Austria); 6 Elba (Italy); 7 Covas de Barroso (Portugal). **Asia**: 8 Zavitinskoe; 9 Malkhan; 10 Orlovka (Russia); 11 Altai; 12 Guanpo; 13 Jaijika; 14 Nanping; 15 Maoantan (China). **Africa**: 16 Khnefissat (Morocco); 17 Kenticha (Ethiopia); 18 Igbeti (Nigeria); 19 Manono (Kongo); 20 Bijyojyo (Rwanda); 21 Kivuvu (Burundi); 22 Alto Ligonha (Mozambique); 23 Betafo (Madagascar); 24 Union; 25 Homestead (South Africa); 26 Rubicon (Namibia). **North America**: 27 Bernic Lake, Cat Lake, Winnipeg River Pegmatites; 28 Little Najanni (Canada); 30 Maine; 31 Massachusetts; 32 Connecticut; 33 North Carolina; 34 Red Ace Wisconsin; 35 Quartz Creek Colorado; 36 Harding New Mexico; 37 San Diego County California (USA). **South America**: 38 Mamoes (Parnamirim); 39 Volta Grande (Minas Gerais) Brazil; 40 Las Cuevas; 41 San Luis (Argentina). **Australia**: 42 Londonderry; 43 Greenbushes; 44 Wodgina (Western Australia)

pegmatite sheets and display dimensions of 100 m in length, 50 m in width and up to 30 m in thickness. The massive pollucite mineralisation of the Tanco LCT pegmatite was discovered by accident during the 1920s in one drill hole intersection and made accessible in the 1950s (Wright 1963; Stilling et al. 2006). Initial reserves were calculated to 125,000–150,000 t at ore grades of 25–30 wt% Cs_2O. Alluvial cassiterite in the Bikita pegmatite field was discovered in 1909. Subsequent mining and exploration for Sn, Be and Li then lead to the exposure of the massive pollucite mineralisation within the Bikita Main Quarry area. At present pollucite is mined from both LCT pegmatite deposits. Bikita Minerals (Private) Ltd. operates an open pit mine while the Tanco LCT pegmatite is mined underground by Tantalum Mining Corporation of Canada Ltd. However, partial collapses in the area of the mines crowning pillar happened in 2010 and 2013, which lead to temporal closures of the operation at Bernic Lake. Mining continued in 2015 while work to stabilise the area was still in progress. The daily mine production was set not to exceed

1000 t (USGS-Cs-2017). Mining of pollucite in the past was reported from the occurrences in Namibia, Sweden and China. The other locations only contain minor amounts of pollucite in small masses, lensoid bodies or pods (<1 m in diameter), as interstitial grains, or crystals in miaroles and fractures. Especially the latter are mined as gemstones.

1.4 General Characteristics and Classification of Pegmatites

Pegmatites have been a long time of interest because of their economic contents of a large number of commodities including Sn, Nb, Ta, Li, Rb, Cs, Be, W, Au, and rare earth elements (REE), as well as phosphates, feldspars, mica and quartz. Furthermore, they are known for their high quality and colourful gemstones and mineral specimens. The term pegmatite is derived from the Greek and means "to bind together", as it first was used to describe the intimate intergrowth of quartz and feldspar within graphic or granophyric granite. Many geological textbooks describe pegmatites as very coarse grained igneous rocks of usually granitic composition that are enriched in rare elements such as Li, Ta, Nb or REE. London (2008) stated that this definition is problematic because it fails to convey the full spectrum of pegmatite textures and proposed as a revision:

> An essential igneous rock, commonly of granitic composition, that is distinguished from other igneous rocks by its extreme coarse but variable grain size, or by an abundance of crystals with skeletal, graphic, or strongly directional growth habit. Pegmatites occur as sharply bounded homogeneous to zoned bodies within igneous or metamorphic host rocks.

London (2008) stated that pegmatites are unambiguously originated from a silicate magma, which includes a silicate melt as liquid phase, crystals and vapour bubbles. Already the early classifications of Jahns (1955) and Schneiderhöhn (1961) subdivided the pegmatites into aqueous, igneous and metamorphic classes. The aqueous class includes lateral secretion and selective solution or aqueous anatexis as principle process. The igneous class includes viscous magmas and highly fluid magmas, even segregated or injected. Furthermore, it includes the source or origin of the pegmatite melt, which was formed by anatexis of pre-existing rocks (Stewart 1978), or represents juvenile discharge from a mantle source. The igneous model further includes interactions with aqueous solutions derived from the magma. In contrast, the metamorphic class explains pegmatite formation due to partial anatexis, recrystallisation and replacement or secretion by metamorphic differentiation.

At present, the most widely used classification of pegmatites as established by Černý and Ercit (2005) is based on the work of Ginsburg et al. (1979). A subdivison into three families considers a combination of emplacement, depth, metamorphic grade, and minor element contents: (I) Lithium–Cesium–Tantalum (LCT) pegmatites; (II) Niobium–Yttrium–Fluorine (NYF) pegmatites and (III) Mixed LCT-NYF pegmatites (Table 1.2). The main advantage of this classification is the inte-

Table 1.2 Family system of petrogenetic classification scheme of granitic pegmatites, after Černý and Ercit (2005)

Family	Pegmatite subclass	Geochemical signature	Pegmatite bulk composition	Granite bulk composition	Granite association	Source rock lithologies
LCT	REL-Li MI-Li	Li, Rb, **Cs**, Be, Sn, Ga, Ta > Nb (B, P, F)	Peraluminous to subaluminous	Peraluminous S, I, mixed S + I types	Synorogenic to late-orogenic (to anorogenic)	Undepleted upper-to-middle crust supracrustals, basement gneisses
NYF	REL-REE MI-REE	Nb > Ta, Ti, Y, Sc, REE, Zr, U, Th, F	Subaluminous to metaluminous (to subalkaline)	Peraluminous to subaluminous, metaluminous A and I types	Syn-, late-, post- to mainly anorogenic, homogeneous	Depleted middle-to-lower crustal granulites, juvenile granitoids
Mixed	Crossbred LCT & NYF	Mixed	Metaluminous to moderately peraluminous	Subaluminous to slightly peraluminous	Postorogenic to anorogenic, heterogeneous	Mixed protoliths, assimilation of supracrustals in NYF granitoids

MI—miarolithic; REL—rare element. Peraluminous—A/CNK > 1; subaluminous—A/CNK ~ 1; metaluminous—A/CNK < 1 at A/NK > 1; subalkaline—A/NK ~ 1; peralkaline—A/NK < 1; A = [Al$_2$O$_3$]; CNK = [CaO + Na$_2$O + K$_2$O]; NK = [Na$_2$O + K$_2$O], in molecular values

gration of the plutonic derivation. Accordingly, LCT pegmatites are suggested to be related to synorogenic, late-orogenic and anorogenic granites of peraluminous with sedimentary (S), metaluminous igneous (I) or mixed S+I type character formed due to melting of undepleted upper to middle-crust or basement gneisses (Černý and Ercit 2005). In contrast, NYF pegmatites are related to predominantly anorogenic granites with varying compositions of anorogenic (A) and I-type character that are derived by melting of depleted middle to lower crustal granulites or juvenile granitoids. The mixed LCT-NYF type pegmatites exhibit characteristics of both. Indeed, this source pluton based classification schemata has major implications for exploration, as it limits the geological environments in which LCT and NYF type pegmatites should occur.

Pegmatites generally occur as groups or clusters of several dozens to >1000 individual pegmatite bodies that not necessarily share a common genetical background. Černý (1982) proposed a subdivison from the large regional to the small local scale into pegmatite province, pegmatite belt, pegmatite district, pegmatite field and pegmatite group. Pegmatites that belong to a group are interpreted to have been derived from a single intrusive body.

In the classical view, pegmatites represent the products of igneous fractionation of granitic melts (Jahns and Burnham 1969; Ginsburg et al. 1979; Jahns 1982; Černý and Ercit 2005; London 2008). Unfortunately, this relation is sometimes not obvious from a regional geological scale. Evidence for a genetic link between granites and pegmatites is found for the Phanerozoic rare-element pegmatites which spatially surround a granite intrusion. However, this relation is not observable for most Archean and Proterozoic rare element pegmatites. These are commonly separated from their igneous sources as they intruded along deep fault systems in tectonically active regimes. Also the genetic relations of the Archean and Proterozoic rare element pegmatites are often hidden due to the present erosional level (Černý 1982; Dill 2015).

Fig. 1.3 Simplified
chemical evolution through a
Li rich pegmatite group with
distance from the granitic
melt source, modified after
Trueman and Černý (1982),
and London (2008)

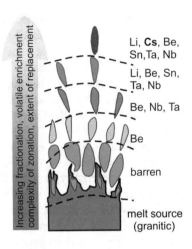

The regional zonation within individual pegmatite groups and pegmatite fields is
another major argument for an igneous derivation of pegmatites. Accordingly, the
pegmatites intruded into the host rocks above and surrounding the upper portions of
a source granite. Continuous fractionation then will cause that the most evolved pegmatites that are enriched in rare elements and volatile components will form farthest
away from a source granite (Fig. 1.3). This tendency is particularly emphasised in
vertical direction, but can be influenced due to structural controls (Černý 1982). The
general zonation sequence in a pegmatite field or group can comprise up to eight
individual zones (Černý 1982; London 2008).

(1) Barren pegmatites of granitic textures, with magnetite and biotite.
(2) Barren plagioclase-microcline pegmatites, in part graphic, with biotite and
    schorl.
(3) Microcline pegmatites, in part graphic, with schorl, muscovite and beryl.
(4) Zoned microcline-albite pegmatites, in part albitised, with muscovite, schorl,
    beryl and Nb–Ta oxides.
(5) Zoned microcline-albite pegmatites, extensively replaced with Li, Rb, Cs, Be,
    Ta, B, P and F mineralisation.
(6) Albite pegmatites with Li, Be, Sn and Ta mineralisation.
(7) Relatively homogenous albite-spodumene pegmatites with minor Be, Ta and Sn
    mineralisation.
(8) Essentially quartz veins with minor feldspar(s) and one or more of beryl, cassi-
    terite and wolframite.

However, this generalised sequence is usually not completely developed in a
single pegmatite field or group. Several different models have been proposed for
this zonality. At present, the general accepted view is the intrusion along a thermal
gradient that surrounds the parental intrusion, with the most fractionated pegmatites
formed at lower temperatures, farther away from the contact to the source granite.

Other models include pulsating injection of different pegmatite types, or a separate ascent of late metasomatic fluids (Černý 1982).

The composition of most pegmatites is comparable to that of granite. Also pegmatites of mafic, intermediate or alkaline composition are known. The principal chemical components of felsic pegmatites are Al_2O_3, SiO_2, Na_2O and K_2O that will form the prevalent minerals quartz, plagioclase and K-feldspar. Rare elements including Li, Cs, Ta, Nb, Sn, Be, P, B and the REE can be enriched to weight percent levels. These elements then collectively can constitute a major portion of the pegmatite and crystallise as mineral assemblages or monomineralic zones that are distinctly non granitic in composition. The zonal structure of pegmatites within a pegmatite field is also reflected in its compositions. These trends are normally recognised by applying element ratios (K/Rb; K/Cs; Rb/Sr; Rb/Tl; Nb/Ta; Al/Ga; and, thus can be used as exploration tool for the identification of possible rare metal mineralisation. Such fractionation trends extend the trends observed in the source granite and are therefore interpreted to serve as petrogenetic indicators (Černý 1982; London 2008; Kaeter et al. 2018).

As the vast majority of the pegmatites have compositions close to that of granites and granodiorites, the essential characteristics to distinguish pegmatites from all other igneous rocks are five different textural attributes (London 2008). These are:

(1) Extremely coarse grain size in relation to igneous rocks of similar compositions.
(2) Extremely variable grain size and grain size distribution, whereas the grain sizes generally increases from the margins to the core of a pegmatite body.
(3) Sharply bounded spatial zonation of mineral assemblages or monomineralic zones.
(4) Frequent graphic intergrowths or skeletal crystal habits as sign of rapid cooling.
(5) A highly anisotropic fabric, emphasised by a strong alignment of non-equidimensional minerals with long axes perpendicular to the margins of the body (e.g., comb structures, stockscheider), as well as layered or radial fabrics.

Pegmatites that contain economic mineralisations display a great diversity of internal zonations and replacement phenomena. As summarised by Cameron et al. (1949) there are:

(1) Zones which form more or less complete concentric shells that follow the shape of the pegmatite sheets and dikes. From the margin inwards, these are subdivided into (a) border zone, (b) wall zone, (c) the intermediate zone with outer, middle, inner and core-margin parts, and the (d) core.
(2) Replacement bodies, which are formed predominantly by replacement of pre-existing units, with or without obvious structural control.
(3) Fracture fillings, which represent generally tabular assemblages filling fractures in preexisting units.

The origin of these zonations is in part related to their different shapes, attitudes and conditions of consolidation. Individual pegmatite bodies can exhibit a great variability of forms and are interpreted to be controlled by the depth of emplacement, mechanical properties of the host rock, the ratio of the internal magma versus the lithostatic pressure, tectonic and metamorphic regime at the time of emplacement and competency of the enclosing host rocks. Pegmatites that are located within their parent granite tend to form schlieren, irregular and bulbous masses or pods that are interpreted to have been formed by the crystallisation from residual melts in situ. Other types of pegmatites that are common within granitic plutons are networks of fracture filling dikes. Although, most of them are barren, they can exhibit some enrichment in Li and Be (Černý 1982).

The sizes of pegmatites can range from tiny millimeter scaled veinlets of incipient leucosome separation within migmatites up to several hundreds of meters. Martin (1964) reported that the dimension of the main Li–Rb–Cs–Be–Ta bearing pegmatite within the Bikita pegmatite field in Zimbabwe is at least 1.8 km and up to 300 m wide in outcrop scale. Even larger pegmatite bodies that cover several tens of km^2 are reported from the Mama granite pegmatite field in the Baikal region (Černý 1982).

1.5 Controls on Pegmatite Formation and Evolution

The most accepted model at present explains pegmatite genesis from residual melts where incompatible components, fluxes, volatiles and rare elements concentrate subsequent to the crystallisation of granitic plutons (Černý 1982; London 2008). Fluxes and volatiles cause a lowering of the crystallisation temperature, decrease melt polymerisation, viscosity and nucleation rates. Furthermore, fluxes and volatiles are interpreted to increase diffusion rates and solubility and thus are considered to be a critical factor that controls the development of large crystals and the vast varieties of pegmatitic textures (Simmons and Webber 2008). Even if many pegmatites exhibit a complex internal structure, they appear to have formed from a single intrusive magmatic event. The structures are thought to reflect a complex internal crystallisation history that results in a large grain size range (<0.1 mm to several 10 m), consistent changes in mineralogy and mineral chemistry from wall to core zones, increasing concentrations of fluxes, volatiles and rare elements. It is generally accepted that pegmatites crystallise from their margins towards their interiors. The fundamentals for this model were introduced by Jahns and Burnham (1969) who suggested that pegmatites evolve from residual granitic melts with coexisting water vapour and silicate melt. Experimental work suggested that pegmatites crystallise in equilibrium of a granite melt and coexisting hydrous fluid at or slightly below the hydrous granite liquidus (Jahns and Burnham 1969). It was recognised that water behaves as an incompatible phase that increases in the residual melt. At the point of H_2O saturation, a discrete water-rich vapour phase is separated. Albeit the fundamentals of this general concept are still undisputed, more recent experiments by London (1992, 2005) indicate that the presence of a hydrous vapour phase is not essential for the development of pegmatite textures during the early crystallisation stages.

Contrasting opinions exist about the emplacement temperature and temperatures of the final stages of pegmatite crystallisation. The model of Jahns and Burnham (1969) proposed that pegmatites crystallise at or near the hydrous minimum melt temperatures of about 600 °C. Two-feldspar thermometry and fluid inclusion studies pointed out that crystallisation of the pegmatite can go down to temperatures as low as 240–270 °C (Anderson et al. 2001). Such low temperatures are interpreted to result from undercooling of the melt due to the effects of volatiles and fluxes. Fluxes such as B, F, P and Li can alter the properties of the melt by lowering the crystallisation temperatures to about 100–300 °C, decrease the viscosity, melt polymerisation and nucleation rates and increase the solubility and diffusion rates of certain elements (Simmons et al. 2003). Increasing solubility and diffusion rates are achieved due to network modifying behaviour of fluxing elements that also prevent or hinder the formation of nuclei. A combination of these two effects facilitates ion migration over greater distances and promotes the growth of only a few nuclei that result in much larger but concurrently fewer crystals (Simmons and Webber 2008).

A long time it was accepted that crystal size in igneous rocks is linked to crystal growth rates. Thus, crystal size can be seen as indicator of the cooling rates, with small crystals growing quickly from a rapidly cooling magma producing a fine grained "granitic" texture. In contrast, large crystals grow slowly from a slowly cooling magma producing a coarse grained phaneritic texture that is by definition one of the most typical characteristics of pegmatites. However, this simple model fails in most pegmatites, as coarse grained units are closely associated with fine grained and in part aplitic units invoking great variability in the cooling history. However, textural relationships of minerals represent an essential tool to estimate the relative degree of melt undercooling, nucleation rate and growth rate (London 2008).

Despite one can estimate relative cooling rates using textural evidences, it was not possible to evaluate the total time that a pegmatite melt needs for cooling. Estimated times range in the order of several thousand to millions of years (Simmons and Webber 2008). Conductive cooling models (Morgan and London 1999) suggested that the time range has to be estimated more likely in days to a couple of months. It is likewise possible to transfer a melt to a metastable and undercooled state by removal of a fluxing component (Simmons and Webber 2008). This chemical quenching is achieved due to incorporation of the fluxing component (B, F) into a crystallising phase as tourmaline or mica. The remaining melt is then not longer fluxed by the removed component and is transferred to an undercooled state.

Numerous attempts to relate pegmatite types or subtypes to magma genesis or tectonic regimes, as already successfully applied for granites, have been tested, but no comparable models were established for pegmatites so far (Wise 1999). Martin and De Vito (2005) argued that a depth zone classification alone cannot account for the two main geochemical categories of LCT and NYF pegmatites. Instead, they propose that the tectonic setting determines the nature of the parent magma and the derivative rare-element enriched magmas. Accordingly, LCT pegmatites are related to compressional tectonic settings (orogenic, calc-alkaline suites) whereas NYF pegmatites form in extensional tectonic setting (anorogenic alkaline suites). Mixed LCT-NYF pegmatites were proposed to result from contamination, either at the magmatic or postmagmatic stage.

Among the diverse factors that are used for the classification and assessment of pegmatites, the geochemical signature and degree of geochemical fractionation is the most crucial parameter with regard to its economic potential. Studied fractionation patterns for LCT pegmatite exploration include LILE as Li, Rb, Cs, or HFSE as Ta, Nb, Zr and U (Trueman and Černý 1982; Černý 1989, 2005; Breaks et al. 2005; Galeschuk and Vanstone 2005).

1.6   Pegmatite Age Distribution and Continental Crust Formation

LCT pegmatite systems are known from each geological era. However, as pegmatites represent derivates of granitic magmatism, they occur more frequent at times of increased continental magmatic activity. Tkachev (2011) demonstrated that increased activity of continental granitic magmatism and LCT pegmatite emplacement occurred in a regular periodicity of 800 Ma. Major peaks at 2650–2600 Ma, 1900–1850 Ma, 1000–950 Ma, and 300–250 Ma of the global age distribution of LCT pegmatite fields can be outlined (Fig. 1.4). Tkachev (2011) further identified subordinate peaks at 2850–2800 Ma, 2100–2050 Ma, 1200–1150 Ma and 550–500 Ma. These age ranges correlate well with an increased continental crust magmatism and are further in a good agreement with the proposed supercontinent cycles of Bradley (2011).

Although continental crust magmatism appears as old as 3850 Ma, there are no LCT pegmatites known prior to about 3150 Ma (Fig. 1.4). Each peak in the times of increased LCT pegmatite emplacement had a relative short time duration of about 50 Ma. Pollucite is known to occur in minor to trace quantities in LCT

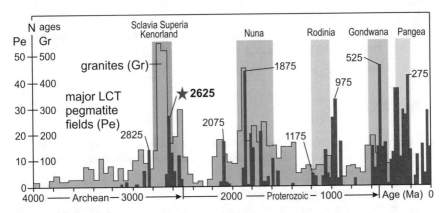

Fig. 1.4 Age distributions of major LCT pegmatite fields Pe and granites Gr (Condie et al. 2009; McCauley and Bradley 2014). Age distribution of major LCT pegmatite fields as determined by Tkachev (2011). Age distribution peaks of the major (bold numbers) and minor pegmatite cycles. The colored bars highlight the approximate assembly times of supercontinents. Red star marks Bikita and Tanco pegmatite fields with massive pollucite mineralisations

pegmatites throughout all these periods. The ages of the known LCT pegmatites that contain economic quantities of massive pollucite mineralisation (Bikita and Tanco LCT pegmatites) coincide with the first major peak of granitic pegmatite formation at 2650–2600 Ma (Fig. 1.4). The age of the Bikita pegmatite was recently updated by Melcher et al. (2015) to be 2617 ± 1 Ma (U–Pb LA-ICPMS on columbite). For the Tanco pegmatite Camacho et al. (2012) reported a Rb–Sr age on lepidolite of 2631 ± 12 Ma, and Melcher et al. (2017) a columbite-cassiterite U–Pb age of 2624 ± 24 Ma. These ages coincide well with the proposed age for the first supercontinent assemblage with the Vaalbara, Superia, Scaliva/Kenorland cratons (Bleeker 2003; McCauley and Bradley 2014). Also most of the presently known large and economic important and several smaller LCT pegmatite deposits are of Meso- to Neo-Archean age (Fig. 1.2; Bradley 2011). The LCT pegmatites at Wodgina (Pilbara Craton, Western Australia), Union (Kaapval Craton, RSA) and Vasin Myl'k (Kola Craton, Russia) are of Meso-Archean age and correlate with the first minor peak of LCT pegmatite formation at 2850–2800 Ma. Numerous Proterozoic, Paleozoic and Mesozoic LCT pegmatites are known, but these do not have the size and economic potential of the Archean ones, including massive pollucite mineralisation. Thus the Neo-Archean time window at 2650–2600 Ma, appears as most prospective, as geological conditions and factors that control the concentration of rare metal concentrations such as Li, Cs, Ta and Sn in pegmatites apparently have been realised there.

The LCT pegmatites in the Archean Yilgarn and Pilbara Cratons in Western Australia match this prospective time window, however, yet no systematic investigation for Cs and massive pollucite mineralisation were performed there. As there are only a few studies on the formation of pollucite in LCT pegmatites and none of these was able to explain the massive Cs mineralisation, the Bikita pegmatites in the Zimbabwe Craton has been investigated for comparison. In Western Australia, a total of 631 rock samples were taken from 19 target areas. The results were condensed to four locations, the Londonderry, Cattlin Creek and Mount Deans pegmatite fields in the Yilgarn Craton, and the pegmatites at Wodgina in the Pilbara Craton.

References

Anderson AJ, Clark AH, Gray S (2001) The occurrence and origin of zabuyelite (Li_2CO_3) in spodumene-hosted fluid inclusions: implications for the internal evolution of rare-element granitic pegmatites. Can Mineral 39:1513–1527

Audi G, Bersillon O, Blachot J, Wapstra AH (2003) The NUBASE evaluation of nuclear and decay properties. Nucl Phys A 729:3–128

Barnes EM, Weis D, Groat LA (2012) Significant Li isotope fractionation in geochemically evolved rare element-bearing pegmatites from the Little Nahanni Pegmatite Group, NWT, Canada. Lithos 132–133:21–36

Barrer RM, McCallum N (1951) Intracrystalline water in pollucite. Nature 167:1071. https://doi.org/10.1038/1671071a0

Bea F, Pereira MD, Corretge LD, Fershtater GB (1994) Differentiation of strongly peraluminous, perphosphorus granites: the Pedrobernardo pluton, central Spain. Geochim Cosmochim Acta 58:2609–2627

Bebout GE, Bebout AE, Graham CM (2007) Cycling of B, Li, and LILE (K, Cs, Rb, Ba, Sr) into subduction zones: SIMS evidence from micas in high-P/T metasedimentary rocks. Chem Geol 239:284–304

Beger MB (1969) The crystal structure and chemical composition of pollucite. Z Kristallog 129:280–302

Bick M, Prinz H, Steinmetz A (2010) Cesium and cesium compounds. Ullmann's encyclopedia of industrial chemistry, 6 pp. https://doi.org/10.1002/14356007.a06_153.pub2

Bleeker W (2003) The late Archean record: a puzzle in ca. 35 pieces. Lithos 71:99–134

Bradley DC (2011) Secular trends in the geologic record and the supercontinent cycle. Earth Sci Rev 108:16–33

Bradley DC, McCauley A, Stillings LM (2017) Mineral-deposit model for Lithium–Cesium–Tantalum pegmatites. USGS Scientific Investigations Report 2010-5070-O, 48 pp. https://pubs.usgs.gov/sir/2010/5070/o/sir20105070o.pdf

Breaks FW, Selway JB, Tindle AG (2005) Fertile peraluminous granites and related rare-element pegmatites, Superior Province of Ontario. In: Linnen RL, Samson IM (eds) Rare-element geochemistry and mineral deposits, vol 17. Geological Association of Canada Short Course Notes, pp 87–125

Breithaupt JFA (1846) Neue Mineralien. 4. and 5. Kastor und Pollux. Annalen der Physik 145–11:429–442. https://doi.org/10.1002/andp.18461451111

Camacho A, Baadsgaard H, Davis D, Černý P (2012) Radiogenic isotope systematics of the Tanco and Silverleaf granitic pegmatites, Winnipeg River pegmatite district, Manitoba. Can Mineral 50:1775–1792

Cameron EN, Jahns RH, McNair A, Page LR (1949) Internal structure of granitic pegmatites. Econ Geol Monogr Ser 2, 115

Černý P (1974) The present status of the analcime-pollucite series. Can Mineral 12:334–341

Černý P (1975) Alkali variations in pegmatitic beryl and their petrogenetic implications. N Jb Mineral Abh 123:198–212

Černý P (1982) Petrogenesis of granitic pegmatites. In: Cerny P (Ed) Short course in granitic pegmatites in science and industry, vol 8. Mineralogical Association of Canada Short Course Handbook, pp 405–461

Černý P (1989) Contrasting geochemistry of two pegmatite fields in Manitoba: products of juvenile Aphebian crust and polycyclic Archean evolution. Precambr Res 45:215–234

Černý P (2005) The Tanco rare-element pegmatite deposit, Manitoba; regional context, internal anatomy, and global comparisons. In: Linnen RL, Samson IM (eds) Rare-element geochemistry and mineral deposits, vol 17. Geological Association of Canada Short Course Notes, pp 127–158

Černý P, Ercit TS (2005) Rubidium and cesium-dominant micas in granitic pegmatites. Can Mineral 43:2005–2026

Černý P, Simpson FM (1977) The Tanco pegmatite at Bernic Lake Manitoba. IX Beryl. Can Mineral 15:489–499

Černý P, Simpson FM (1978) The Tanco pegmatite at Bernic Lake, Manitoba; X pollucite. Can Mineral 16:325–333

Černý P, Meintzer RE, Anderson AJ (1985) Extreme fractionation in rare element granitic pegmatites: selected examples of data and mechanisms. Can Mineral 23:381–421

Černý P, Chapman R, Teertstra DK, Novák M (2003) Rubidium- and cesium-dominant micas in granitic pegmatites. Am Mineral 88:1832–1835

Condie KC, Belousova E, Griffin KL, Sircombe KN (2009) Granitoid events in space and time: constraints from igneous and detrital zircon age spectra. Gondwana Res 15:228–242

Dailey SR, Christiansen EH, Dorais MJ, Kowallis BJ, Fernandez DP, Johnson DM (2018) Origin of the fluorine- and beryllium-rich rhyolites of the Spor Mountain Formation. Am Miner 103:1228–1252

Dill HG (2015) Pegmatites and aplites: their genetic and applied ore geology. Ore Geol Rev 69:417–561

Ellis AJ, Mahon WAJ (1977) Chemistry and hydrothermal systems. Academic Press, 392 pp

Endo A, Yoshikawa E, Muramatsu N, Takizawa N, Kawai T, Unuma H, Sasaki A, Masano A, Takeyamac Y, Kaharac T (2013) The removal of cesium ion with natural Itaya zeolite and the ion exchange characteristics. J Chem Technol Biotechnol 88:1597–1602

Galeschuk CR, Vanstone PJ (2005) Exploration for buried rare-element pegmatites in the Bernic Lake area of southern Manitoba. In: Linnen RL, Samson IM (eds) Rare-element geochemistry and mineral deposits. Geological Association of Canada Short Course Notes, vol 17, pp 159–173

Gatta GD, Rinaldi R, McIntyre GJ, Nenert G, Bellatreccia F, Guastoni A, Della Ventura G (2009) On the crystal structure and crystal chemistry of pollucite, (Cs, Na)16Al16Si32O96 × nH2O: a natural microporous material of interest in nuclear technology. Am Mineral 94:1560–1568

Ginsburg AI, Timofeyev LN, Feldman LG (1979) Principles of geology of the granitic pegmatites. Nedra, Moscow (CCCP), 296 pp (in Russian)

Hall A, Jarvis KE, Walsh JN (1993) The variation of cesium and 37 other elements in the Sardinian granite batholith, and the significance of cesium for granite petrogenesis. Contrib Miner Petrol 114:160–170

Hart SR, Reid MR (1991) Rb/Cs fractionation: a link between granulite metamorphism and the S-process. Geochim Cosmochim Acta 55:2379–2383

Horstman EL (1957) The distribution of lithium, rubidium, and caesium in igneous and sedimentary rocks. Geochim Cosmochim Acta 12:1–28

Icenhower J, London D (1995) An experimental study of element partitioning among biotite, muscovite, and coexisting peraluminous silicic melt at 200 MPa (H$_2$O). Am Mineral 80:1229–1251

Jahns RH (1955) The study of pegmatites. In: Bateman AM (Ed) Economic geology 50th anniversary, vol 1905–1955, pp 1025–1130

Jahns RH (1982) Internal evolution of pegmatite bodies. In: Cerný P (Ed) Short course in granitic pegmatites in science and industry, vol 8. Mineralogical Association of Canada Short Course Handbook, pp 293–327

Jahns RH, Burnham CW (1969) Experimental studies of pegmatite genesis: I. A model for the derivation and crystallization of granitic pegmatites. Econ Geol 64:843–864

Jambor J, Vanko DA (1990) New mineral names. Am Mineral 75:706–713

Jing Z, Hao W, He X, Fan J, Zhang Y, Miao J, Jin F (2016) A novel hydrothermal method to convert incineration ash intopollucite for the immobilization of a simulant radioactive cesium. J Hazard Mater 306:220–229

Jing Z, Cai K, Li Y, Fan J, Zhang Y, Miao J, Chen Y, Jin F (2017) Hydrothermal synthesis of pollucite, analcime and their solid solutions and analysis of their properties. J Nucl Mater 488:63–69

Kaeter D, Barros R, Menuge JF, Chew DM (2018) The magmatic-hydrothermal transition in rare-element pegmatites from southeast Ireland: LA-ICP-MS chemical mapping of muscovite and columbite-tantalite. Geochim Cosmochim Acta 240:98–130

Kamiya N, Nishi K, Yokomori Y (2008) Crystal structure of pollucite. Z Kristallog 223:584–590

Keith TEC, Thompson JM, Mazs RE (1983) Selective concentration of cesium in analcime during hydrothermal alteration, Yellowstone National Park, Wyoming. Geochim Cosmochim Acta 47:795–804

Kirchhoff GR, Bunsen RWE (1861) Chemische analyse durch spectralbeobachtungen. Annalen der Physik und Chemie 189:337–381. https://doi.org/10.1002/andp.18611890702

Kobayashi H, Yanase I, Mitamura T (1997) A new model for the pollucite thermal expansion mechanism. J Am Ceram Soc 80:2161–2164

Kretz R (1983) Symbols for rock forming minerals. Am Mineral 68:277–279

Lagache M, Dujon SC, Sebastian A (1995) Assemblages of Li–Cs pegmatite minerals in equilibrium with a fluid from their primary crystallization until their hydrothermal alteration: an experimental study. Miner Petrol 55:131–143

London D (1990) Internal differentiation of rare element pegmatites; a synthesis of recent research. Geol Soc Am Spec Pap 246:35–50. https://doi.org/10.1130/SPE246-p35

London D (1992) The application of experimental petrology to the genesis and crystallization of granitic pegmatites. Can Mineral 30:499–540

London D (1995) Geochemical features of peraluminous granites, pegmatites, and rhyolites as sources of lithophile metal deposits. In: Thompson JFH (Ed) Magmas, fluids, and ore deposits, vol 23. Mineralogical Association of Canada, Short Course Handbook, pp 175–202

London D (2005) Geochemistry of alkali and alkaline earth elements in ore-forming granites, pegmatites, and rhyolites. In: Linnen RL, Samson IM (eds) Rare-element geochemistry and mineral deposits, vol 17. Geological Association of Canada Short Course, pp 175–199

London D (2008) Pegmatites. Spec Publ Can Mineral 10:368 pp. ISBN: 978-0-921294-47-4

London D, Morgan GBVI, Icenhower J (1998) Stability and solubility of pollucite in the granite system at 200 MPa H_2O. Can Mineral 36:497–510

Lyubetskaya T, Korenaga J (2007) Chemical composition of Earth's primitive mantle and its variance: 1. Method and results. J Geophys Res 112(B03211), 21 pp. https://doi.org/10.1029/2005jb004223

Martin HJ (1964) The Bikita Tinfield. Southern Rhodesia Geol Surv Bull 58:114–131

Martin RF, De Vito C (2005) The patterns of enrichment in felsic pegmatites ultimately depend on tectonic setting. Can Minerar 43:2027–2048

McCauley A, Bradley DC (2014) The global age distribution of granitic pegmatites. Can Mineral 52:183–190

McDonough WF, Sun SS, Ringwood AE, Jagoutz E, Hofmann AW (1992) Potassium, rubidium, and cesium in the Earth and Moon and the evolution of the mantle of the Earth. Geochim Cosmochim Acta 56:1001–1012

Melcher F, Graupner T, Gäbler HE, Sitnikova M, Henjes-Kunst F, Oberthuer T, Gerdes A, Dewaele S (2015) Tantalum-(niobium-tin) mineralisation in African pegmatites and rare metal granites: constraints from Ta-Nb oxide mineralogy, geochemistry and U-Pb geochronology. Ore Geol Rev 64:667–719. https://doi.org/10.1016/j.oregeorev.2013.09.003

Melcher F, Graupner T, Gäbler HE, Sitnikova M, Henjes-Kunst F, Oberthür T, Gerdes A, Badanina E, Chudy T (2017) Mineralogical and chemical evolution of tantalum-(niobium-tin) mineralisation in pegmatites and granites. Part 2: worldwide examples (excluding Africa) and an overview of global metallogenetic patterns. Ore Geol Rev 89:946–987. https://doi.org/10.1016/j.oregeorev.2016.03.014

Melzer S, Wunder B (2001) K-Rb-Cs partitioning between phlogopite and fluid: experiments and consequences for the LILE signatures of island arc basalts. Lithos 59:69–90

Merefield JR, Brice CJ, Palmer AJ (1981) Caesium from former Dartmoor volcanism: its incorporation in New Red sediments of SW England. J Geol Soc London 138:145–152

Morgan GB, London D (1999) Crystallization of the little three layered pegmatite-aplite dike, Ramona District, California. Contrib Miner Petrol 136:310–330

Naray-Szabo S (1938) Die Struktur des Pollucits. Z Kristallog 99:277–282

Neßler J, Seifert T, Gutzmer J, Müller A (2017) Beitrag zur Erkundung und metallogenetischen Charakteristik der Li-Sn-W-Greisenlagerstätte Zinnwald, Osterzgebirge, Deutschland. Freiberger Forschungshefte C 552 - Geowissenschaften, TU Bergakademie Freiberg, Germany: 390 pp and Appendix (summary and list of figures and tables in English)

Newnham RE (1967) Crystal structure and optical properties of pollucite. Am Mineral 52:1515–1518

Osichkina RG (2006) Regularities of trace element distribution in water-salt systems as indicators of the genesis of potassium salt rocks: an example from the Upper Jurassic Halogen Formation of Central Asia. Geochem Int 44:164–174

Pinsky C, Bose R, Taylor JR, McKee JSC, Lapointe C, Birchall J (1981) Cesium in mammals: acute toxicity, organ changes and tissue accumulation. J Environ Sci Health, Part A 16:549–567

Plattner CF (1846) Chemische Untersuchung zweier neuen, vom Herrn Prof. Breithaupt mineralogisch bestimmten Mineralien von der Insel Elba. Ann Phys 145:443–447. https://doi.org/10.1002/andp.18461451112

Potter EG, Taylor RP, Jones PC, Lalonde AE, Pearse GHK, Rowe R (2009) Sokolovaite and evolved lithian micas from the Eastern Moblan granitic pegmatite, Opatica subprovince, Quebec, Canada. Can Mineral 47:337–349

Redkin HF, Hemley JJ (2000) Experimental Cs and Sr sorption on analcime in rockbuffered systems at 250–300 °C and Psat and the thermodynamic evaluation of mineral solubilities and phase relations. Eur J Miner 12:999–1014

Roselieb K, Jambon A (1997) Tracer diffusion of potassium, rubidium, and cesium in a supercooled jadeite melt. Geochim Cosmochim Acta 61:3101–3110

Rudnick RL, Gao S (2014) 4.1—composition of the continental crust. In: Holland HD, Turekian KK (eds) Treatise on geochemistry (2nd ed), pp 1–51

Schneiderhöhn H (1961) Die Erzlagerstätten der Erde; Band II, Die Pegmatite. Gustav Fischer Verlag Stuttgart, 720 pp

Sebastian A, Lagache M (1990) Experimental study of the equilibrium between pollucite, albite and hydrothermal fluid in pegmatitic systems. Min Mag 54:447–454

Seifert T (2008) Metallogeny and petrogenesis of lamprophyres in the mid-European variscides: post-collisional magmatism and its relationship to late-Variscan ore forming processes (Bohemian Massif). IOS Press BV, Amsterdam, Netherlands, 303 pp

Selway JB, Breaks FW, Tindle AG (2005) A review of rare-element (Li-Cs-Ta) pegmatite exploration techniques for the Superior Province, Canada, and large worldwide Tantalum deposits. Explor Min Geol 114:1–30

Simmons WB, Webber KL (2008) Pegmatite genesis: state of the art. Eur J Miner 20:421–438

Simmons WB, Webber KL, Falster AU, Nizamoff JW (2003) Pegmatology—pegmatite mineralogy, petrology and petrogenesis. Rubellite Press New Orleans, 176 pp

Smeds SA, Černý P (1989) Pollucite from the proterozoic petalite bearing pegmatites of Utö, Stockholm archipelago, Sweden. Geol Fören Stockh Förh 111:361–372

Stewart DB (1978) Petrogenesis of lithium-rich pegmatites. Am Mineral 63:970–980

Stilling A, Černý P, Vanstone PJ (2006) The Tanco pegmatite at Bernic Lake, Manitoba; XVI, Zonal and bulk compositions and their petrogenetic significance. Can Mineral 44:599–623

Taylor SR, McLennan SM (1985) The continental crust; its composition and evolution: an examination of the geochemical record preserved in sedimentary rocks. Blackwell Oxford, 312 pp

Taylor WH (1930) The structure of analcite ($NaAlSi_2O_6 \times H_2O$). Z Kristallog 74:1–19

Teertstra DK, Černý P (1995) First natural occurrence of end-member pollucite: a product of low-temperature reequilibration. Eur J Miner 7:1137–1148

Teertstra DK, Černý P, Chapman R (1992) Compositional heterogeneity of pollucite from high grade Dyke, Maskwa Lake, southeastern Manitoba. Can Mineral 30:687–697

Teertstra DK, Lahti SI, Alviola R, Černý P (1993) Pollucite and its alteration in Finnish pegmatites. Geol S Finl Bull 368, 42

Teertstra DK, Černý P, Langhof J, Smeds SA, Grensman F (1996) Pollucite in Sweden: occurrences, crystal chemistry, petrology and subsolidus history. GFF 118:141–149

Teertstra DK, Černý P, Novák M (1995) Compositional and textural evolution of pollucite in pegmatites of the Moldanubicum. Mineral Petrol 55:37–51

Thomas R, Davidson P (2012) Water in granite and pegmatite-forming melts. Ore Geol Rev 46:32–46

Tindle AG, Selway JB, Breaks FW (2005) Liddicoatite and associated species from the McCombe spodumene-subtype rare-element granitic pegmatite. Can Mineral 43:769–793

Tkachev AV (2011) Evolution of metallogeny of granitic pegmatites associated with orogens through geological time. In: Sial AN, Bettencourt JS, de Campos CP, Ferreira VP (eds) Granite-related ore deposits. Geological Society, London, Special Publications, vol 350, pp 7–23

Trueman DL, Černý P (1982) Exploration for rare-element granitic pegmatites. In: Cerny P (Ed) Short course in granitic pegmatites in science and industry, vol 8. Mineralogical Association of Canada Short Course Handbook, pp 463–494

USGS-Cs-2017 (2017) U.S. Geological Survey Mineral Commodity Summaries (MCS). https://minerals.usgs.gov/minerals/pubs/mcs/2017/mcs2017.pdf

Wahlberg JS, Fishman MJ (1962) Adsorption of cesium on clay minerals. US Geol Surv Bull 1140A, 30 pp

Wang RC, Hu H, Zhang AC, Fontan H, Zhang H, De Parseval P, Jiang SY (2007) Cs-dominant polylithionite in the Koktokay #3 pegmatite, Altai, NW China: in situ micro-characterization and implication for the storage of radioactive cesium. Contrib Mineral Petrol 153(3):355–367. https://doi.org/10.1007/s00410-006-0151-y

Watson EB (1979) Diffusion of cesium ions in H_2O-saturated granitic melt. Science 205(4412):1259–1260. https://doi.org/10.1126/science.205.4412.1259

Wedepohl KH (1978) Handbook of geochemistry, vol II/5. Springer, Berlin, 1546 pp

Wise MA (1999) Characterization and classification of NYF-type pegmatites. Can Mineral 37:802–803

Wood SA, Williams-Jones AE (1993) Theoretical studies of the alteration of spodumene, petalite, eucryptite and pollucite in granitic pegmatites: exchange reactions with alkali feldspars. Contrib Miner Petrol 114:255–263

Wright CM (1963) Geology and origin of the pollucite-bearing Montgary pegmatite, Manitoba. Geol Soc Am Bull 74:919–946

Xiao Y, Lavis S, Niu Y, Pearce JA, Li H, Wang H, Davidson J (2012) Trace-element transport during subduction-zone ultrahigh-pressure metamorphism: evidence from western Tianshan, China. Geol Soc Am Bull 124:1113–1129

Yanase I, Kobayashi H, Shibasaki Y, Mitamura T (1997) Tetragonal-to-cubic structural phase transition in pollucite by low-temperature X-ray powder diffraction. J Am Ceram Soc 80:2693–2695

Yokomori Y, Asazuki K, Kamiya N, Yano Y, Akamatsu K, Toda T, Aruga A, Kaneo Y, Matsuoka S, Nishi K, Matsumoto S (2014) Final storage of radioactive cesium by pollucite hydrothermal synthesis. Scientific Rep 4, 4195 pp. https://doi.org/10.1038/srep04195

Chapter 2
Geological Settings of Archean Rare-Metal Pegmatites

2.1 Zimbabwe Craton

As they host the LCT pegmatites, the understanding of the geological evolution of Archean Cratons is of exceptional importance. That applies especially for the massive pollucite mineralisation in the LCT pegmatite district of Bikita, in the Masvingo greenstone belt in the SE part of the Zimbabwe Craton (Fig. 2.1). The Zimbabwe Craton is bounded on the southern side by the granulite-facies Northern Margin Zone of the Limpopo belt and to the north by the Zambezi belt. Both represent Neo-Archean mobile belts accreted onto the Zimbabwe Craton at 2700–2600 Ma (Rollinson and Whitehouse 2011).

The Archean stratigraphy of the Zimbabwe Craton is subdivided into three groups. These are (I) the 3600–3000 Ma Sebakwaian Group; (II) the 3100–2950 Ma Bulawayan Group and (III) the 2700–2600 Ma Shamvaian Group (Dodson et al. 2001). The Sebakwaian Group comprise mafic to ultramafic metavolcanic and metasedimentary rocks which are overlain by banded iron stones and metapelites. This succession of an extensional basin was later intruded by various granitoid suites as the 3380 Ma Mushandike granite, and the 3350 Ma Mont d'Or granite (Taylor et al. 1991; Dodson et al. 2001).

Separated by a prominent unconformity, the Lower Bulawayan Group (3100–2950 Ma) overlies the Sebakwaian Group. This succession comprises a transgressive assemblage of conglomerates, quartz-arenites, siltstones, shale, cherts and ironstones. Occurrence of ultramafic and mafic lavas within the upper portion of the succession is indicative for an extensional regime within a rift. Subsequent felsic magmatism started at 2900 Ma and lead to the deposition of andesites, dacites, rhyolites and their volcanoclastic equivalents (Wilson et al. 1995; Jelsma and Dirks 2002). Furthermore, volcanic rocks are intercalated with cherts, ironstones and are accompanied by the emplacement of major komatiites and high Mg- and tholeiitic basalts. Tonalite-trondhjemite-granodiorite type granitoids (TTG) of the Chingezi

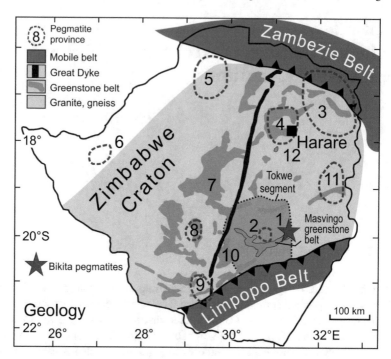

Fig. 2.1 Zimbabwe Craton with the Masvingo greenstone belt. LCT pegmatite occurences in the craton are: 1 Bikita; 2 Masvingo; 3 NE of Mutoko; 4 Harare; 5 Karoi district; 6 Kamativi; 7 Que Que; 8 N of Bulawayo; 9 SE of Bulawayo; 10 Mweza Range; 11 W of Mutare; 12 Hwedza district; modified after Grubb (1985), Horstwood et al. (1999), Dodson et al. (2001) and Oberthuer et al. (2002)

Suite intruded this succession throughout their development from 2980 to 2750 Ma (Taylor et al. 1991; Horstwood et al. 1999) and are interpreted as plutonic equivalents of the Lower Bulawayan felsic volcanic rocks. During a subsequent time period from 2750 to 2580 Ma most of the present greenstone belt succession was formed in an extensive period of mafic to ultramafic volcanism. The volcanism was concomitant with the eruption of vast volumes of basalts and komatiites, and associated volcanoclastic counterparts. These rocks are summarised as Upper Bulawayan Group (Wilson et al. 1978; Wilson 1990) and are interpreted as depositions within rifts on a preexisting continental basement.

The period of predominantly mafic volcanism was followed by a phase of felsic volcanism. Clastic and volcanoclastic sedimentary rocks of this stage rest unconformably upon the Upper Bulawayan Group and are assigned to the Shamavian Group. Its deposition was accompanied by the emplacement of the Sesombi and Weda Suite tonalites and granodiorites at about 2700–2690 Ma. This event is interpreted as mantle plume activity and a concomitant vertical overturn of crust (Dirks and Jelsma 1998; Prendergast 2004; Shimizu et al. 2005).

A dramatic tectonic change occurred at about 2670 Ma in the southern part of the Zimbabwe Craton (Fig. 2.1). The Limpopo Belt started moving northwards, resulting in extensive deformation and metamorphism up to granulite-facies conditions within the Northern Limpopo Thrust Zone (Wilson 1990; Frei et al. 1999). The compressional event was accompanied by granulite-facies metamorphism and partial melting that lead to granitoid emplacement in the middle and upper crust (Rollinson and Whitehouse 2011). In the Zimbabwe Craton this event is expressed by the intrusion of the 2601 Ma Chilimanzi Suite of granites. Nd and Sr whole rock isotope analysis (Jelsma and Dirks 2002) could show that the Chilimanzi Suite magmas were derived from older continental crust and not accompanied by the addition of larger portions of magmas from the mantle. This "recycling" of older continental crust is an unique feature of the 2670–2590 Ma event and may signal a major change in the tectonic structure of the Earth (Kusky 1998). As the previous magmatic events are interpreted to result from vertical tectonics and associated overturn of the crust, the 2610–2590 Ma magmatism is interpreted to be induced by horizontal tectonics within a modern-style subduction setting. Ages of LCT pegmatite emplacement at Bikita with 2617 ± 1 Ma, and at Benson with 2587 ± 4 Ma (Melcher et al. 2015) coincided with the emplacement of this magmatic suite. Later, a significant addition of material to the crust was due to multiple intrusions of mafic to ultramafic dike swarms. The most significant event was the intrusion of the Great Dike at 2575 Ma that crosscuts the Zimbabwe Craton in N–S direction (Oberthuer et al. 2002; Soederlund et al. 2010).

2.1.1 Pegmatites in the Zimbabwe Craton

Pegmatites in the Zimbabwe Craton are grouped into 12 provinces (Fig. 2.1). The oldest pegmatites so far known in Zimbabwe, occur in the Kwekwe area and are Paleoarchean (3370 Ma; Grubb 1985). The majority of the pegmatites in the Bikita, Masvingo, Harare and Kamativi areas are Meso- to Neo-Archean and also Mesoproterozoic in age (Grubb 1985; Melcher et al. 2015). The youngest pegmatites with Cambrian to Devonian ages (540–400 Ma) are found in the Karoi district and in the area NE of Mutoko (Grubb 1985). Based on their mineralisation characteristic, host rock association and age, Grubb (1985) subdivided the pegmatites in Zimbabwe into the Miami-, Kamativi- and Bikita-types. Among the 12 pegmatite provinces in Zimbabwe, at present only the pegmatites within the Bikita (Martin 1964) and Kamativi areas, as well as from the Benson pegmatite field within the Motoko area are important for mining of Li, Ta, Sn, Nb and Cs. Information on the geological setting of the other pegmatites is rare, and most of them probably are important for gemstone mining. Until present, the Bikita pegmatite field is the sole occurrence on the entire Zimbabwe Craton that is known to host massive pollucite mineralisation (Martin 1964).

2.1.2 The Bikita LCT Pegmatite Field in the Masvingo Greenstone Belt

The Bikita LCT pegmatite field is located within the easternmost portion of the Masvingo greenstone belt. It consists of a northern part with N–S trending and a southern part with E–W trending pegmatites. Individual pegmatites are hosted within predominantly greenstone belt lithology with metabasalts and metadiorites of the Upper Bulawayan Group and only minor within pelitic rocks that are assigned by Wilson (1964) to the Lower Bulawayan Group. The greenstone belts are rimmed by older granitic gneisses that probably belong to the > 3300 Ma Sebakwaian Group and two granite plutons of the 2600 Ma Chilimanzi Suite (Gwavava and Ranganai 2009). These are the Chikwanda pluton to the N and the Charumbira pluton along the southern margin of the greenstone belt.

The Masvingo greenstone belt is an ENE-trending, about 100 km long and 22 km wide belt (Fig. 2.1) in the southern Zimbabwe Craton, and located about 50 km NW to the tectonic boundary with the Northern Marginal Zone of the Limpopo belt (Dodson et al. 2001; Prendergast and Wingate 2007; Gwavava and Ranganai 2009). Detailed mapping of the Masvingo greenstone belt and its surrounding granitoids and gneisses was performed by Wilson (1964). Martin (1964) conducted a detailed mapping and description of the "Bikita Tinfield". The main part of the Masvingo greenstone belt was formed by the Lower and Upper Bulawayan Group and comprises quartz-mica schist, pelitic metasedimentary rocks with intercalated bands or lenses of banded ironstone, banded quartzite or limestone. The mafic rocks of the Upper Bulawayan Group within the greenstone belt are metabasalts, chlorite-, or hornblende-schists and metadiorites.

During the NW directed thrusting of the Limpopo belt onto the Zimbabwe Craton at 2700 Ma this rock succession was metamorphosed and tightly folded to a synclinal structure. Subsequent extension started at 2670 Ma and was concomitant with the emplacement of the Chilimanzi Suite granitoids. Gwavava and Ranganai (2009) unravelled the relation between the greenstone belt and adjacent granites by geophysical methods. The contact of the granites to the greenstone belt is almost vertical, whereas the contact to the older gneisses to the S is very shallow and domal in shape. Extension enabled the formation of the NW–SE striking Gono fault and the N–S striking Popoteke fault in the Masvingo greenstone belt (Fig. 2.2a). The Gono fault displays some degree of dextral shearing that is associated with the formation of a second set of N–S striking normal faults. This structural setting favoured the emplacement of pegmatite melts of the Bikita pegmatite province along the N–S striking faults. The pegmatite melts are generally accepted to represent highly fractionated residual melts derived from the crystallisation of the close-by younger granites of the Chilimanzi Suite or the Sesombi Suite (Martin 1964; Grubb 1985; Melcher et al. 2015). At present it is not possible to link the pegmatites of the Bikita province to individual granite intrusions. Numerous dolerite dikes intruded with various sizes and shapes, and penetrated the complete stratigraphic succession (Fig. 2.2a, b). At least some of these dikes are interpreted as offshoots from the N–S trending

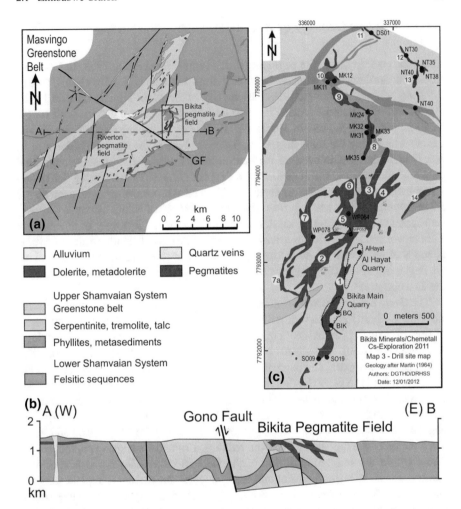

Fig. 2.2 Geological setting of the Bikita pegmatite field, modified after Martin (1964) and Wilson (1964). **a** Eastern part of Masvingo greenstone belt, with pegmatite fields N (Bikita) and S (Riverton) of the Gono fault (GF). **b** Section through the eastern portion of the Masvingo greenstone belt with position of Bikita pegmatites. **c** Tenement of the Bikita Minerals (Private) Ltd. with individual pegmatite sheets and sampling sites. Single pegmatites (P.) in the Bikita field: 1 Main P.; 2 Spodumene P.; 3 Boomerang P.; 4 East P.; 5 Shaft P.; 6 West P.; 7 Giant East P.; 7a Giant West P.; 8 Mauve Kop South P.; 9 Mauve Kop Middle P.; 10 Mauve Kop North P.; 11 Dam Site P.; 12 Nigel North P.; 13 Nigel South P.; 14 Ranga P

Great Dike intrusion complex that were emplaced about 100 km W of the Masvingo greenstone belt at 2575 Ma (Oberthuer et al. 2002). Furthermore, several other generations of dolerite dikes and quartz porphyry dikes are present, although outcrops of the latter are restricted to the surrounding granites. The last remarkable event within the Masvingo greenstone belt is the formation of several generations of quartz veins.

2.1.3 Deposit Geology of Bikita

Cassiterite within the Bikita pegmatite district was discovered in 1909. In 1911, the Koestlich Brothers pegged the Sequenula claims, now the Nigel Tin claims. At this time, tantalite was discovered about 10 km SW of the Main Tinfield (Martin 1964). Production of cassiterite started in 1916 and continued, even though with only a small output, until the end of World War II. Due to increasing demand for Be and Li minerals during the following years, Bikita evolved to the leading producer of Li ore. Petalite and lepidolite were the main minerals produced with some additional output of spodumene, amblygonite and eucryptite (Martin 1964). During this period mining was active within the Al Hayat and Bikita Tin block claims, which are on the Main Pegmatite Field and still form part of the holdings of Bikita Minerals (Private) Ltd. Another prospecting boom started in 1958, after the discovery of emeralds in Zimbabwe, and lead to the discovery of several of dozens of smaller pegmatite veins N and S of the centre of the Bikita pegmatite district. Mining for Rb, Li and since the beginning of the 1990s also increasingly for Cs continued until the beginning of the 21st century.

The Gono fault separates the northern Bikita pegmatite field in the hanging wall, from the southern Riverton pegmatite field, situated in the footwall. Both fields consist of several dozens of individual pegmatite bodies emplaced along a 15–20 km long structure (Fig. 2.2c). The Riverton field is characterised by Be-bearing pegmatites at its easternmost part. The pegmatites evolve into Be- and Ta-bearing types in the centre, and then to Sn- and Li-dominated types in the westernmost portions. A similar trend was observed within the Bikita field. Beryl bearing pegmatites are predominant within the northernmost portions about 20 km N of the Bikita Main Quarry. The pegmatites become more Ta and Sn pronounced at about the Mauve Kop and Nigel Tin locations. The highest evolved and thus Li-, Rb- and Cs-bearing pegmatites are found within the southernmost part of the field, in the Al Hayat and Bikita Main Quarry areas. Such geochemical and mineralogical zonations are typical for LCT pegmatites. The N–S trend of the Bikita pegmatite field suggest a possible source towards the N. This area is characterised by a low gravity anomaly that suggests a hidden granite intrusion. No data on age and geochemical composition of the granites in the vicinity of Bikita are available at present. Therefore, the suspected source granite of Bikita pegmatites may belong to the Charumbira or Chikwanda Suite of granites.

The field work in 2011 for the study presented here was conducted only on the license area of Bikita Minerals Ltd. starting from the Mauve Kop location in the N towards the South Office location S of the Bikita Main Quarry (Fig. 2.2c). The area is predominantly underlain by the Upper Bulawayan Group greenstone belt metabasalts and metadiorites, and subordinate by metapelites that are assigned by Martin (1964) to belong to the Sebakwaian Group (Fig. 2.2b). No outcrops of granite were discovered during the fieldwork (Dittrich 2016; Dittrich et al. 2015). In the first geological map by Martin (1964), the pegmatite field is subdivided into 15 localities (Fig. 2.2c). The predominant bodies crop out within the southern portion of the

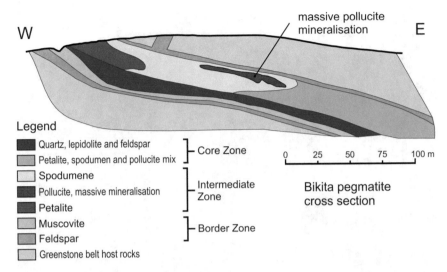

Fig. 2.3 Schematic section through the Bikita Quarry pegmatite with its mineral zones, modified after Cooper (1964)

pegmatite field. It consists of a N–S trending, ~2 km long pegmatite sheet and includes the Bikita Main Quarry and the Al Hayat Quarry that are currently being mined for pollucite and petalite, respectively. According to Martin (1964), this pegmatite or body was emplaced within the crest of an anticlinal structure at the eastern end of the deposit that plunges into a synclinal structure on the W-side. Consequently, the pegmatite sheet dips gently towards the E on the eastern portion and gently towards the W in the western portion. The Bikita Main Quarry and Al Hayat Quarry are located along the eastern limb of the anticlinal structure, whereas the Giant West and Spodumene pegmatite localities are on its western limb (Fig. 2.2c). This anticlinal structure appears to straighten towards the N and vanishes within the outcrop areas of the East-, Boomerang-, Shaft- and West-pegmatite. Another characteristic of the southern portion of the pegmatite field is the occurrence of serpentinite lenses in close spatial association to the pegmatite.

The Bikita main pegmatite is about 30–40 m in total thickness and dips with 30–45° towards the E (Cooper 1964). In general, the pegmatite exhibits three individual zones. The border zone consists of a feldspar and a muscovite dominated zone. It is followed by an intermediate zone consisting of predominantly feldspar and domains that host either petalite or spodumene as major Li minerals. The massive pollucite mineralisation rest as almost monomineralic ore bodies within the spodumene dominated domains (Fig. 2.3). Individual ore bodies are lens to sheet like in shape and have dimensions of approximately 100 × 100 m with up to 15 m in total thickness. In 2011, two lenses of massive pollucite mineralisation cropped out within the workings of the Bikita Main Quarry. A second pegmatite sheet crops out further to the N. This pegmatite sheet trends NNW–SSE and comprises the Mauve

Fig. 2.4 a Bikita Main Quarry, view to the N, with the contact to the hanging wall greenstone belt and the location of the two sampled massive pollucite mineralisation ore bodies BQ and BIK. **b** View on the E wall of the Bikita Main Quarry with the BIK ore body. **c** Fine network of lepidolite veins in the massive pollucite. **d** Dam Site pegmatite viewed to the N with small (2 × 3 m) lens of massive pollucite mineralisation. **e** Contact of a dolerite dike that crosscuts the Mauve Kop Central pegmatite

Kop South, Middle and North pegmatites (Fig. 2.4). A prominent dolerite dike cross-cuts the Mauve Kop North pegmatite. Several smaller pegmatites occur further E of the Mauve Kop pegmatites. These are the Nigel Tin and Dam Site pegmatites. In 2011 and 2012, one small pollucite lens was cropping out in small exploration workings at the Dam Site locality.

2.2   Yilgarn Craton

The Yilgarn Craton represents one of the largest structural units of the Earth's crust that are entirely made up of Archean (> 2500 Ma) rocks (Fig. 2.5). Based on structural work (Cassidy et al. 2006; Blewett et al. 2010), the Yilgarn Craton is divided into an eastern and western part along the Waroonga fault in the N and the Ida fault in the S. The Yilgarn Craton is separated into six terranes (Fig. 2.6). Each of these terranes is further subdivided into smaller tectonic units like the Coolgardie Domain, Norseman Domain or Southern Cross Domain. The Cattlin Creek LCT pegmatite deposits form part of the Southern Cross Terrane of the West Yilgarn Craton. The Londonderry and Mount Deans pegmatite fields form part of the Kalgoorlie Terrane.

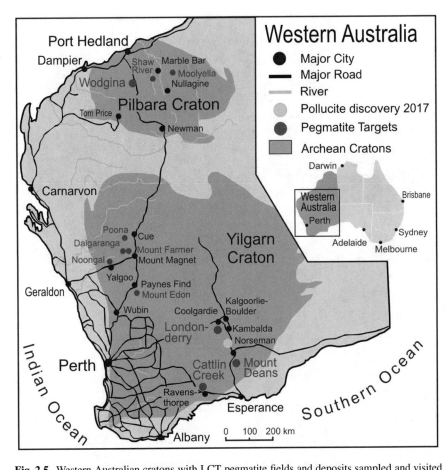

Fig. 2.5 Western Australian cratons with LCT pegmatite fields and deposits sampled and visited for this study in 2011 and 2012. Green point marks recent pollucite discovery (Poineer Resources Ltd. 2017; Crook 2018)

Due to its importance for Ni and especially Au mining, many detailed geological investigations were performed within the Eastern Yilgarn Craton. New data are available also from the Western Yilgarn Craton (Goscombe et al. 2009). The characteristic feature of the Yilgarn Craton are greenstone belts that are enclosed by vast areas of granitic and gneissic rocks. The granite-gneiss complexes characteristically comprise about 70–90% of the outcropping rocks and are made up of multiple stacks of intrusions throughout the Archean terrain (Cassidy et al. 2002). Greenstone belts are typically composed of a series of metamorphic mafic to ultramafic rocks (e.g., metabasalts, komatiites) or their plutonic equivalents (metagabbro), and associated metasedimentary (metapelites) or volcanosedimentary rocks (Smithies et al. 2018). The greenstone belts are equally located in the Eastern and Western Yilgarn Craton. The dimensions of the greenstone belts range from several tens up to several

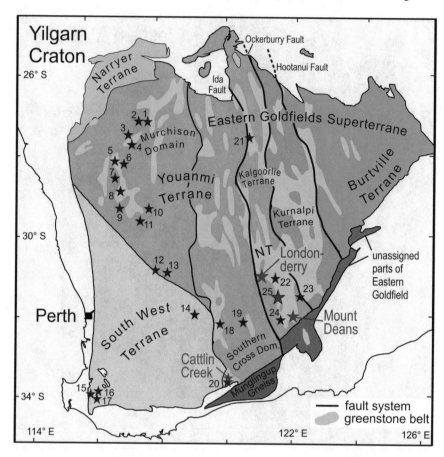

Fig. 2.6 Tectonic subdivison of the Yilgran Craton into terranes and domains with Norseman Terrane (NT). Red stars indicate pegmatite fields considered for this study. Blue stars indicate other pegmatite fields and locations: 1 Coodardy: 2 Poona; 3 Warda Warra; 4 Dalgaranga-Mount Farmer; 5 Melville (Noongal); 6 Edah Hill; 7 Badja Homestead barite pegmatite; 8 Mount Mulgine (Pickano Springs); 9 Rothsay (Seleka) beryl pegmatite; 10 Mount Edon (Pynes Find; Goodingnow); 11 Mount Gibson Kaolinite pegmatite; 12 Waddouring Rock (Hill) allanite pegmatites; 13 Mukinbudin; 14 Holleton; 15 Ferndale Estate pegmatite; 16 Greenbushes; 17 Smithfield, 18 Mount Holland, 19 Mount Day, 20 Cocanarup, 21 Kathleen Valley, 22 Mount Marion; 23 Binnerringie; 24 Mount Thirsty: 25 Pollucite discovery in the Sinclair Cesium Project (Pioneer Resources Ltd. 2017; Crook 2018). Map is modified after Cassidy et al. (2006) and Jacobson et al. (2007)

hundreds of kilometres in length and several kilometres up to 50 km in width. Their internal structure is characterised by a network of interconnected shear zones, separating lenticular high-strain domains from lower-strain domains and displacing them against granites and gneisses (Goscombe et al. 2009).

2.2.1 Terranes in the Yilgarn Craton

The Kalgoorlie Terrane in the Eastern Yilgarn Craton (Fig. 2.6) is about 600 km long and bordered by large-scale shear zones that penetrated through the crust and probably into the lithospheric mantle (Goscombe et al. 2009). The oldest rocks (> 2780 Ma) within the Kalgoorlie Terrane are found within the Wiluna Domain. The base of this stratigraphic succession consists of alternating felsic and komatiitic rocks followed by intermediate volcanic rocks. The subsequent period was characterised by the emplacement of large volumes of tholeiitic and high-MgO basalts followed by the sedimentation of turbidites (Kositcin et al. 2008). Goscombe et al. (2009) interpreted that the Wiluna Domain was formed in the period from 2810 to 2785 Ma. The only age data available from this greenstone belt strata are from younger intermediate to felsic dikes that crosscut the rocks of the Wiluna Domain. They have ages of 2749 ± 7 Ma (U-Pb zircon, Kent and Hagemann 1996) and 2770 ± 3 Ma (U-Pb zircon; Dunphy et al. 2003).

The time after the sedimentation of this volcanosedimentary sequence (~2785 Ma) is not recorded within the greenstone belt stratigraphy of the Eastern Goldfield Superterrane. The next younger stratigraphic sequence (2715–2695 Ma) is found within the Kambalda Sequence (Goscombe et al. 2009). The lithostratigraphy of Kambalda Sequence is subdivided into the Lunnon Basalt, Kambalda Komatiite and the Paringa Basalt. Also minor rhyolites and gabbro intrusions are observed. The age of the Kambalda Komatiite was determined at 2706 ± 36 Ma (Re-Os; Foster et al. (1996), and confirmed by a 2708 ± 7 Ma U-Pb zircon age (Nelson 1997) from an intercalated rhyolite succession.

The Kambalda Sequence is unconformably overlain by the Kalgoorlie Sequence and concomitant with a major change in volcanism from mafic-ultramafic to predominantly felsic melt compositions. The base of the Kalgoorlie Sequence is defined by the Spargoville Formation that represents a thick succession of dacitic to rhyolitic volcanic rocks deposited in an arc-adjacent volcano-bound basin with dacite breccias (2686 ± 3 Ma) and dacite lavas (2698 ± 6 Ma). The Spargoville Formation is unconformably overlain by the deep marine Black Flag Formation with volcanoclastic rocks of rhyolitic to dacitic provenance and minor felsic (granodiorite, tonalite) and mafic (dolerite, gabbro) intrusive rocks. Age constraints for the deposition of the Black Flag Formation range from about 2708 Ma to 2658 ± 3 Ma. The younger Kurrawang Formation comprises metasedimentary rocks in an upward fining succession of coarse grained quartzofeldspatholithic conglomerates, sandstone and mudrock. Although, no direct age information is yet available, it is accepted that the Kurrawang Formation was deposited after 2657 Ma (Kositcin et al. 2008; Krapež and Hand 2008; Goscombe et al. 2009).

The Norseman Terrane is interpreted by some authors as a structural subdomain of the Kalgoorlie Terrane. The rocks of the Norseman Terrane belong to two distinct stratigraphic cycles. The first cycle from about 2810–2785 Ma represents the Lower Norseman Domain (Kositcin et al. 2008). A second stratigraphic record between 2730 and 2695 Ma is further referred to as Upper Norseman Domain. Both domains are separated by an unconformity and a stratigraphic hiatus covering at least 50 Ma. The U-Pb zircon age dating in a meta-rhyolite yielded 2980 ± 4 Ma (Nelson 1997). The Upper Norseman Domain consists of a basal unit of metasedimentary rocks (turbidites, sandstone, chert) that belong to the Noganyer Formation. The U-Pb age determinations by SHRIMP on detrital zircon grains, as well as hydrothermal zircon rims and grains produced ages of 3670–3650 Ma and 2706 ± 5 Ma, respectively (Campbell and Hill 1988). A synvolcanic dolerite sill gave a U-Pb age of 2714 ± 5 Ma (Hill et al. 1989).

Although belonging to the Southern Cross Domain of the Youanmi Terrane, the tectonic units within the Ravensthorpe area are interpreted by some authors as a distinct terrane. The geochemical compositions of the Manyutup tonalite and Annabelle volcanic rocks represent an island arc setting and may have been originated by melting of subducted oceanic plate or metasomatised asthenospheric mantle (Witt 1999). SHRIMP U-Pb age determinations on zircons from felsic intrusive (tonalite, tonalite porphyry) and calc-alkaline volcanic rocks yield ages from 2970 to 2980 Ma (Savage et al. 1995).

The stratigraphic successions of the Yilgarn Craton represent three different cycles of crust formation during the Archean. The Ravensthorpe Terrane is the oldest Terrane and represents a Meso-Archean period from about 2990 to 2955 Ma. In contrast, the two terranes of the Eastern Yilgarn Craton (Kalgoorlie Terrane and Norseman Terrane) are Neo-Archean in age and record two distinct cycles of crustal growth. The older cycle from about 2810 to 2785 Ma is recorded by the Wiluna Domain in the Kalgoorlie Terrane and the Lower Norseman Domain in the Norseman Terrane. The younger cycle of crustal growth started at 2720 Ma and is represented by the Upper Norseman Domain, and the Kambalda Sequence, Kalgoorlie Sequence and late basins in the Kalgoorlie Terranes. In the Norseman Terrane the stratigraphic record ends at about 2695 Ma. The youngest rocks within the Kalgoorlie Terrane are represented by the 2660–2655 Ma Kurrawang Formation.

2.2.2 Tectono-Magmatic Evolution

For the Yilgarn Craton, there are two main contrasting models for the Archean evolution which enclose seven deformational events (D0–D6). The asymmetric model of Goscombe et al. (2009) is based on modern-style plate tectonics starting already in the Meso- and Neo-Archean Eras and suggests a general westward directed subduction type scenario with successive amalgamation of crustal units. In contrast, the amalgamation of the crustal units is explained by van Kranendonk et al. (2007a) only by vertical movements within the crust, induced by constantly upwelling of hot mantle material (Fig. A2).

D0: > 2780 Ma: For the beginning of the crustal evolution of the Eastern Yilgarn Craton Goscombe et al. (2009) recognised two probable scenarios. The asymmetric scenario is based on volcanic arc-type setting with the development of a magmatic

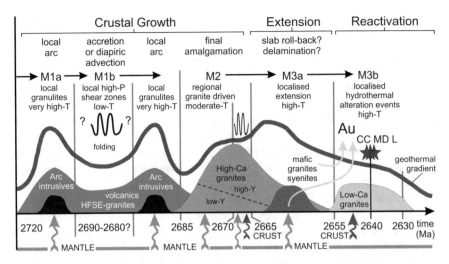

Fig. 2.7 Tectono-magmatic evolution of the Yilgarn Craton between 2720 and 2630 Ma. Metamorphic events are M1a to M3b, modified from Goscombe et al. (2009). Geothermal gradient varies from high to intermediate temperatures. Intrusion ages of various granitic suites with contributions of mantle (green arrows) and crust melts (red arrows); ages of gold mineralisations (Au), and intrusion ages of LCT pegmatites of Cattlin Creek (CC), Mount Deans (ML) and Londonderry (L) are indicated

arc and a related back arc basin. In contrast, the symmetric model is expressed by mantle plume related volcanism (Fig. 2.7).

D1: **2720–2685 Ma**: This time period was characterised by ENE extensional and compressional events in complex magmatic arc, back-arc and mantle plume related environments and was accompanied by emplacement of bimodal magmatic and volcanic rocks, and the Kambalda Komatiite. During this event, most of the precursors of now intensely deformed greenstone belts were emplaced to upper crustal levels (Fig. 2.7). Peak metamorphic conditions at 2700–2685 Ma reached ~730 °C at low pressures between 2.5 and 5.0 kbar, but is only recognised in a few localities. Due to high heat flow from magmatic arcs, this metamorphism was associated to a very high temperature/depth ratio between 45 and 80 °C/km (Goscombe et al. 2009). This granulite-facies metamorphism M1a was active contemporaneous with the deposition of basalts and komatiites which were only affected by low-grade seafloor alteration and regional metamorphism. The domain boundaries such as the Ida and Ockerberry fault systems display evidence for high-P (7.0–8.7 kbar), high-T (570–640 °C), and a low temperature/depth ratio (18–23 °C/km). This localised amphibolite-facies metamorphism is referred to as M1b metamorphism.

High-field-strength-element granites (HFSE-granites) in the Eastern Yilgarn Craton cause a first magmatic peak at 2700–2680 Ma (Champion and Cassidy 2007). Mafic and syenite granitoids are closely associated with Au deposits and possibly related to the transfer of Au from a mantle source into upper crustal levels (Groves and Phillips 1987; Bucci et al. 2004; Bierlein et al. 2006). High-Ca granitoids predominantly crop out within the Kalgoorlie and Kurnalpi Terranes and were emplaced over a long period of time from 2700 to 2640 Ma. Two episodes of High-Ca magmatism within the Kalgoorlie Terrane can be recognised. The first episode with low Y contents started with felsic volcanism at about 2700 Ma, increases to a peak at ~2675 Ma and diminishes until 2670 Ma (Champion and Cassidy 2007). A second episode of High-Ca magmatism with high Y contents emplacement peaks at about 2660 Ma. The tectonic setting for the early phase of magmatism is interpreted to be subduction related at a continental margin and includes a component of crustal recycling. In contrast, the second episode of High-Ca melts exhibits higher Y contents, indicating that no garnet was present in the restite. It may result from melting at shallower crustal levels and involving a higher share of crustal melt (Champion and Sheraton 1997; Cassidy et al. 2002).

D2: 2670–2665 Ma: The extensive volcanism active during D1 terminated at 2670 Ma and the prevalent tectonic regime changed from ENE extension towards NNW contraction. This event is accompanied by upright folding and thrusting and concomitant with regionally extensive metamorphism labelled as M2 that ranges from sub-greenschist- to mid-amphibolite-facies. Peak metamorphic pressures range from 3.5–5.0 kbar with temperatures ranging from 350 to 620 °C. The temperature/depth ratio was elevated at 30–40 °C/km. This thermal peak is interpreted to result from: (a) conduction and advection of heat due to the emplacement of large volumes (60–65%) of High-Ca granitoids, and (b) radiogenic heat from the anomalous high Th-contents produced by the HFSE-type granites emplaced during D1. Main driving force for D2 in the alternative scenarios is a considerable crustal shortening in E–W direction. The asymmetric plate tectonic model explains the crustal shorting due to continuous westward subduction and the formation of an accretionary prism. In contrast, the vertical-style tectonic model suggests mantle plume related convergences and a repeated cycle of crustal overturn. Within the asymmetric model the cause for the pervasive granite plutonism is explained by lower crust delamination induced by subduction of an oceanic plateau followed by a slab roll-back (Blewett et al. 2010). Such a slab roll back may have initiated a change in tectonic setting from contraction (D2) to extension (D3), leading into the M3a metamorphic period (Goscombe et al. 2009).

D3: 2665–2655 Ma: The D3 event represents an extensional regime separated into two subevents. The D3a event dated at 2665–2660 Ma is represented by extensional doming and the development of late basins. The D3b event is contemporaneous to D3a and related to development of crustal-scale extensional shear zones and corresponding basins. This extension is interpreted to cause decompressional melting in the lower crust and the formation of Low-Ca granitoids that constitute about 30% of the granites within the Eastern Yilgarn Craton (Fig. 2.8). Low-Ca biotite granites and granodiorites are interpreted to post-date crustal growth processes such as arc magmatism and volcanism. Their ages range from 2660 to 2610 Ma. The Low-Ca melts were generated by partial melting of a middle to lower crust and do not contain a mantle component, did not involve subduction processes, and are interpreted to be associated with lithospheric extension (Champion and Cassidy 2007; Goscombe et al. 2009).

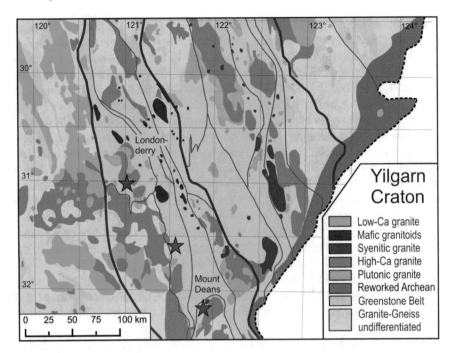

Fig. 2.8 Distribution of various granite groups in the Eastern Yilgarn Craton, and position of Londonderry and Mount Deans pegmatite fields, modified from Goscombe et al. (2009). Pollucite discovery in the Sinclair Cesium Project (Pioneer Resources Ltd. 2017; Crook 2018) is marked with a green star

Peak metamorphic conditions were reached at 500–580 °C and 4 kbar, indicating a high temperature/depth ratio of 40–50 °C and are referred to as M3a metamorphism. The M3a metamorphism and lithospheric extension led to the formation of three new fluid sources and reservoirs. These are (a) metamorphic fluid released by dehydration of hydrous minerals; (b) accumulation of large volumes of saline fluids in new sediment basins with 10–15 km depth; and (c) invasion of dry CO_2-rich and H_2-CH_4-rich fluids with Au from the mantle into the upper crust. Indeed, the M3a metamorphic event coincides with the age of the second Au-mineralisation event in the Eastern Yilgarn Craton (Goscombe et al. 2009).

D4: 2655–2650 Ma: This event can be subdivided into D4a structures with NNW upright folding and reverse faulting, and D4b with NNW sinistral transpression and thrusting (Goscombe et al. 2009). The D4 event is accompanied by the onset of M3b metamorphism (2650–2620 Ma) which is characterised by a change from dip-slip extension towards multiple strike-slip kinematic episodes. This rotation of the extensional regime was the last major change within the geotectonic setting of the Eastern Yilgarn Craton. The 2655–2650 Ma event of D4 is accompanied by intrusion of high-Y granites and some syenites. The M3b metamorphism during D4 is recognisable across most parts of the Eastern Yilgarn Craton and occurred over a wide range of temperatures and pressures (250–500 °C; 3.0–3.5 kbar) indicating a moderately elevated thermal gradient of 30–50 °C/km (Goscombe et al. 2009).

D5: **2650–2630 Ma**: The D5 event is dominated by dextral strike-slip transtension under still ongoing M3b-style metamorphism and the emplacement of the Low-Ca granitic melts formed during D3 into shallower crustal levels (Goscombe et al. 2009). The emplacement of the Low-Ca granitoids led to a short lived increase of the temperature/depth ration of > 70 °C/km, and again is related with the hydrothermal alteration and Au-mineralisation in second- and third-order shear and fault zones (Groves and Phillips 1987; Hagemann and Cassidy 2000; Bucci et al. 2004; Bierlein et al. 2006; Witt et al. 2017, 2018). Between 2645 and 2630 Ma rapid exhumation stripped about 7–10 km of crust indicating another major change in tectonic setting. This tectonic event is interpreted by Goscombe et al. (2009) to be caused by a switch from continental growth to lithospheric thinning after the final assembly of a supercontinent. The termination of the supercontinent assembly (Kenorland; Pehrsson et al. 2013; Griffin et al. 2014) is interpreted to have resulted due to the slab role back and the initiation of M3a lithospheric thinning at 2665 Ma.

D6: **2630 Ma**: The tectonic development of the Eastern Yilgarn Craton is terminated at around 2630 Ma accompanied by minor vertical shortening under still M3b-style tectonic regime and the fading of the Low-Ca type magmatism and Au-mineralisation.

Afterwards, the Yilgarn Craton acted as rigid block and was only locally affected by the intrusion of Paleo- to Mesoproterozoic dikes and a slight overprint in the southern part caused by the amalgamation of the Albany-Fraser-Orogen during the Mesoproterozoic (Dawson et al. 2003; Scibiorskia et al. 2015; Smithies et al. 2015; Stark et al. 2018).

2.2.3 Pegmatites in the Yilgarn Craton

Pegmatites are reported from at least 125 locations within the Yilgarn Craton (Jacobson et al. 2007; Sweetapple 2017). According to their spatial distribution, the pegmatites can be roughly classified into four interregional provinces and geotectonic positions (Fig. 2.6). These are in the Murchison Domain, the Eastern Goldfields Super terrane with the Londonderry and Mount Deans pegmatite fields, the terrane boundary between the Youanmi and South West Terranes with the Cattlin Creek LCT pegmatite, and the South West Terrane with the Greenbushes LCT pegmatite deposit. Greenbushes represents the world largest known reserves of lithium minerals (Partington et al. 1995). According to Talison Lithium Company, at 2012 total proven and probable mineral reserves are 31.5 Mt @ 3.1 wt% Li_2O, and total measured and indicated minerals resources comprise 70.4 Mt @ 2.6 wt% Li_2O.

Most of the LCT pegmatites in the Yilgarn Craton are hosted by mafic to ultramafic volcanic and volcanosedimentary greenstone belt sequences that rest adjacent to vast granite-gneiss complexes (Fig. 2.6). Numerous exploration programs and mining operations for feldspar, Be, Li, Sn and Ta (Londonderry pegmatite field, Bald Hill pegmatite deposit; Jacobson et al. 2007) contributed to information on the geological and genetically setting of the pegmatites. Comparable geological settings of other LCT pegmatite provinces worldwide (Bikita, Tanco) support the potential for the discovery of new deposits in the Yilgarn Craton that may contain economic quantities of massive pollucite mineralisation.

2.2.4 Geological Setting of the Londonderry Pegmatite Field

The Londonderry pegmatite field in the Coolgardie Domain is located about 25 km SW of Coolgardie and about 550 km E of Perth. It is easily accessible via the Great Eastern Highway and the Coolgardie Nepean Mine road (Fig. 2.9a). Small roads and tracks, unnamed and not easily visible from the Coolgardie Nepean Mine road lead to the individual pegmatites. The Londonderry pegmatite field is situated in the Kangaroo Hills greenstone belt which forms an about 20 km long and NE–SW striking structure. It has the typical dome and keel structure of Archean granite greenstone belt terranes, displaying tightly folded, steeply dipping metabasalts and metagabbros, as well as metakomatiites (Fig. 2.9b). The greenstone belt underwent metamorphism of lower- to upper-amphibolite-facies conditions (Mikucki and Roberts 2004). Numerous gneissic and granitoid rocks surround the Kangaroo Hills greenstone belt. Most of the granites intruded post-folding (Hunter 1993).

The pegmatite field comprises at least four known pegmatite sheets: (a) Londonderry Feldspar Quarry; (b) Lepidolite Hill; (c) Tantalum Hill, and (d) Bon Ami pegmatite. The pegmatites at Londonderry were discovered in the early 1900s, when settlers prospected the area for gold. Following a mining period for alluvial tantalite and cassiterite, a first quarry operation for microcline feldspar started at the Feldspar Quarry pegmatite in 1929. Subsequently, discontinued mining produced about 68,000 t of microcline, 1200 t of petalite, 0.84 t of columbite-tantalite and 190 t of beryl (Jacobson et al. 2007). From 1964 to 1965, Western Mining Corporation undertook a detailed exploration program for Li-ore (petalite and/or lepidolite) covering the whole area of the Londonderry pegmatite field. They performed a detailed geological surface mapping and RC drilling which could prove an estimated reserve of 3700 t of petalite in the area of the Lepidolite Hill open pit. This was later mined together with lepidolite and beryl between 1971 and 1973. This exploration attempt reveals no minable reserves of petalite and lepidolite at the Tantalite Hill prospect and Londonderry Feldspar Quarry pegmatite. Mining activities at the Londonderry Feldspar Quarry pegmatite continued with intervals until 1993. Main product was feldspar with subordinate output of beryl, petalite and minor columbite-tantalite. In 1995, Commercial Minerals Ltd. drilled seven holes to identify additional reserves of ceramic grade feldspar. According to Jacobson et al. (2007), a total of 20,000 t "A-grade feldspar" and 50,000 t "B-grade feldspar" reserves could be proven. However, high mica contents within microcline rated these reserves as uneconomic. In 2000, Copper Mines and Metals Ltd. undertook an additional drilling program for Ta at the Lepidolite Hill open pit pegmatite. They performed four vertical RC drill holes

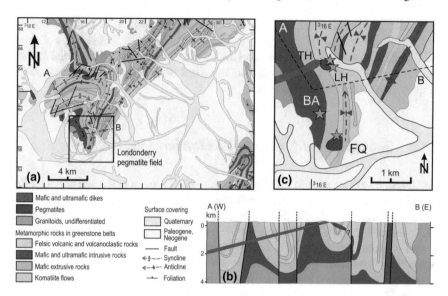

Fig. 2.9 **a** Eastern part of the Kangoroo Hills greenstone belt and geological setting of the Londonderry pegmatite field, modified after Hunter (1988) and Dittrich (2016). **b** Cross section, after Hunter (1988). **c** Single pegmatite bodies in (**a**); BA Bon Ami prospect; FQ Feldspar Quarry pegmatite; LH Lepidolite Hill pegmatite; TH Tantalite Hill pegmatite

and could demonstrate that the pegmatite is up to 15 m in thickness and contains up 82 g/t Ta_2O_5 and 445 ppm Li, which are still uneconomic (Jacobson et al. 2007). The Londonderry pegmatite field was up to 2017, the sole locality in Australia from which pollucite was reported. This pollucite was intersected in one drill hole within the Lepidolite Hill pegmatite area during the 1965 drilling campaign of Western Mining Corporation. The white glassy pollucite that contains a framework of grey mica veinlets was obtained from 24 m depth and likely marked with the sample number "3464/67". Today, this sample is treasured in the Western Australia Museum in Perth as specimen number MDC 4096. An uncalibrated EDS analysis of this sample yielded only half of the nominal Cs_2O of pollucite (Jacobson et al. 2007). Despite this single occurrence of pollucite, the size of the pegmatite field, the general internal zonation and mineralogical associations at Londonderry are comparable to the massive pollucite mineralisation at Bikita (Zimbabwe) and Tanco (Canada). Thus, the Londonderry pegmatite group represents a crucial target for the understanding of the genesis of pollucite bearing LCT pegmatite systems in the Yilgarn Craton.

The Londonderry pegmatite field covers an area of about 5 km^2 and crops out along N–S trending anticlines and synclines within the close proximity of an almost oblique NNW–SSW trending fault (Fig. 2.9b, c). Folding was tight as indicated by an almost vertical foliation of the pegmatite host rocks. Main host for the pegmatites are the metamorphosed komatiite flows. Towards the W and S, the Kangaroo Hill greenstone belt is confined by granitic and gneissic rocks of unknown affinity. Both Archean units, the greenstone belt terrane and the post-collisional granites, are crosscut by an

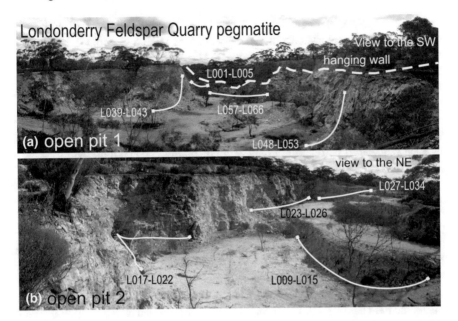

Fig. 2.10 **a** Open pit 1 of the Londonderry Feldspar Quarry pegmatite with view toward SW. Sampling locations along transects are indicated. **b** Open pit 2 with sampling sites, in a view toward NE

ENE–WSW striking Proterozoic mafic to ultramafic dike that may be linked either to the Celebration or the Randalls dike systems (Hunter 1988, 1993). This Proterozoic dike is spatially related with the Tantalum Hill pegmatite. Rb-Sr dating by Turek (1966) indicate an age of 2420 ± 30 Ma for the Celebration dike. Other analyses near Kambalda (Roddick 1974) and Queen Victoria Rocks yielded Rb-Sr ages of 2085 and 2043 Ma, respectively.

The **Londonderry Feldspar Quarry Pegmatite** is the most prominent pegmatite sheet within the Londonderry pegmatite field. This NNE–SSW striking and NW dipping pegmatite sheet is accessible via two about 150 m long, 80–120 m wide and up to 30 m deep open pits (Fig. 2.10). Jacobson et al. (2007) reported a total length of about 1000 m with a thickness ranging from 15 to 66 m and that the complete pegmatite profile consists of at least 7–8 zones (Table A3). The petalite bearing zones are only found within the western quarry and are extensively altered.

The **Lepidolite Hill Pegmatite** is situated about 1.5 km NNW of the Londonderry Feldspar Quarry pegmatite. The pegmatite crops out in two distinct bodies, a NE pegmatite sheet accessible within a 100 m long, 30 m wide and 20 m deep open pit, and a southeastern pegmatite body, that should form a L-shaped outcrop with dips to the NE and NW. The pegmatite contains at least six different zones (Jacobson et al. 2007). Based on the reported thickness and assuming an ideal symmetrical zonation, one can conclude a 60 m of total thickness of the Lepidolite Hill pegmatite, which is comparable to the Bikita and Tanco pegmatites.

The **Tantalite Hill Pegmatite** is barely recognisable at the surface and crops out within small exploration pits (10 m long, 5 m wide and about 2 m deep) in the northernmost part of the pegmatite field. The pegmatite is zoned and contains a border zone of albite-quartz and garnet, a wall zone of albite-quartz and four discontinuous intermediate zones of quartz-albite-microcline, albite-quartz-zinnwaldite, albite-quartz-lepidolite and quartz-microcline-petalite.

From the **Bon Ami Pegmatite** only small outcrops and exploration pits are found about 1 km NW of the Londonderry Feldspar Quarry pegmatite and was described as a flat lying sheet composed of quartz, albite, microcline and muscovite.

2.2.5   Geological Setting of the Mount Deans Pegmatite Field

The Mount Deans pegmatite field is located about 15 km S of the small town of Norseman and about 560 km E of Perth (Figs. 2.6 and 2.8). It is accessible via the Norseman-Esperance Highway and small unpaved tracks. As the small tracks are unnamed they are not easily visible from the Norseman-Esperance Highway. The pegmatite field has an E–W extension of about 3–4 km and a N–S extension of about 6 km and comprises at least 71 individual pegmatite sheets (Jacobson et al. 2007). Pegmatites outcrops from this area were first reported in 1896. Between 1965 and 1967, 7.18 t of cassiterite were mined from the Mount Deans area, most probably from alluvial material (Jacobson et al. 2007). Subsequent work included several exploration attempts for Sn and Ta. Even though drilling identified resources of about 9.1 Mt of ore grading 216 g/t Ta_2O_5, the pegmatite field is still classified as uneconomic.

The Mount Deans pegmatite field is situated within the southern portion of the Kalgoorlie Terrane, sometimes labelled as the Norseman Terrane (Fig. 2.8). The Penneshaw Formation represents the deepest structural level within the greenstone belt of the Norseman Terrane and consists predominantly of amphibolites and massive pillowed basalts, with minor felsic volcanic and sedimentary rocks. This succession is overlain by the Noganyer Formation, characterised by persistent layers of banded iron formation within clastic sedimentary rocks. Several gabbro sills are observed. Both formations were subjected to strong and pervasive deformation accompanied by up to upper-amphibolite-facies metamorphism. The next stratigraphic formation is the Woolyeenyer Formation, a monotonous crustal unit composed of massive basalt flows and a subordinated ultramafic component (Groenewald et al. 2000). Geochronological studies by Nelson (1995) and Campbell and Hill (1988) on a rhyolite from the Penneshaw Formation yielded ages of 2930 ± 4 Ma and 2938 ± 10 Ma. Furthermore, phenocrysts from this rhyolites yield ages as old as 3450 Ma (Hill et al. 1989). Consequently, the temporal geotectonic evolution within the Norseman Terrane differs from that of the Kalgoorlie Terrane (2700 Ma) in the N, as well as from the Southern Cross Domains towards the W.

The Mount Deans pegmatite field crops out alongside the mountain range of Mount Deans. The pegmatites are hosted by metabasalts of the Woolyeenyer Forma-

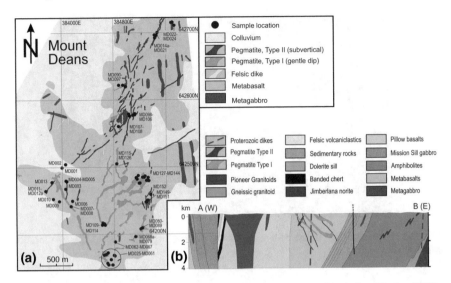

Fig. 2.11 a Mount Deans pegmatite field with sampling locations, modified after Eliyahu (2003).
b Schematic cross section, modified after McGoldrick (1994)

tion and some metagabbroic dikes (Fig. 2.11). A group I of pegmatites consists of up
to 20 m thick gently dipping (20–30°) sheets that exhibit in part a distinct zonation
and in general have a more diverse composition than that of group II (Fig. 2.12).
Minerals within group I are quartz, feldspar (in part cleavelandite), lepidolite (in part
massive zones or ball-like structures), zinnwaldite, garnet, cassiterite and Ta-Nb-
minerals. Jacobson et al. (2007) further reported the occurrence of the Li minerals
petalite and spodumene. In contrast, group II consists of relatively thin (up to 3 m)
and steeply dipping (70–90°) pegmatites which are dominated by quartz and lepidolite with only minor feldspar. The crosscutting relationship indicate that the group I
pegmatites are slightly older than group II.

2.2.6 Geological Setting of the Cattlin Creek Pegmatite Deposit

The Cattlin Creek pegmatite is situated at the SE margin of the Yilgarn Craton within
the Archean Ravensthorpe Terrane that forms the southernmost portion of the Southern Cross Domain (Fig. 2.6). This portion of the Yilgarn Craton is ascribed to the
Youanmi Terrane (Blewett et al. 2010). It consists of Archean greenstone belts and
granites, which are dated at about 3000–2950 Ma (Myers 1993). Based on stratigraphic and geochronological data, Wyche et al. (2012) compared the development
of the Southern Cross Domain with that of the Murchison Terrane in the north.

Fig. 2.12 Exploration pit of the Mount Deans flat dipping pegmatite type 1. Sample location MD075, view to the N

The Ravensthorpe greenstone belt is 75 × 50 km in size and forms one of the southernmost Archean portions of the Southern Cross Domain. Witt (1999) sub-divided the greenstone belt into three tectonic units (a) Ravensthorpe Terrane, (b) Carlingup Terrane, and (c) Cocanarup Terrane. The terranes are separated by thrust faults and comprise metabasalts and metakomatiites, as well as sequences of metased-imentary rocks with intercalated banded iron formations. Geochronological studies on zircon (U-Pb, SHRIMP) yielded ages that range from 3000 to 2950 Ma (Savage et al. 1995; Witt 1999). Beside the Cattlin Creek pegmatite, the Ravensthorpe green-stone belt is the host of the Cocanarup pegmatite field, located at about 20 km SW of Ravensthorpe. The Ravensthorpe Terrane represents a syncline that consists of the Annabelle Volcanic rocks and the Manyutup Tonalite Formation. The Annabelle Volcanic rocks represent a metavolcanic succession dated at 2989 ± 11 Ma (U-Pb SHRIMP; Witt 1999). The Manyutup Tonalite Formation with plutonic rocks that range from quartz-dioritic to granodioritic compositions yielded the same age at 2989 ± 7 Ma (U-Pb SHRIMP; Witt 1999), which indicates a genetic relationship.

The Cattlin Creek pegmatite deposit is located a few kilometres W of the small town Ravensthorpe, about 550 km ESE of Perth and is easily accessible via the South Coast Highway and the Newdegate Ravensthorpe road. The Cattlin Creek pegmatite intruded along NNW–SSE striking thrust faults and was emplaced along the contact between the Manyutup Tonalite and Annabelle Volcanic rocks (Witt 1999). Although already discovered in 1900, the first systematic description of the pegmatite was conducted by Ellis (1944). Sofoulis (1958) proposed a zonation that comprises three internal zones and six mineral assemblages summarised in Table A4. The pegmatites at Cattlin Creek are classified as complex rare element type pegmatites and host significant Li and minor Ta resources (Galaxy Resources Limited 2017). Spodumene was first discovered in small outcrops within the creek bed of the Cattlin Creek at the beginning of the twentieth century. Subsequent exploration

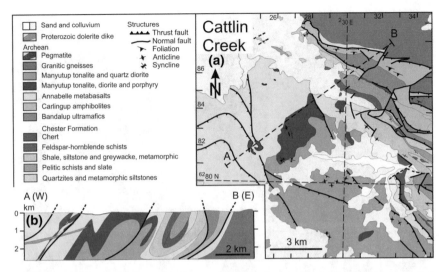

Fig. 2.13 **a** Geology of the Cattlin Creek pegmatite field in the Ravensthorpe greenstone belt, modified after Witt (1996). **b** Cross section, strongly modified after Thom and Lipple (1977)

attempts by various companies included detailed drilling campaigns for Ta and Li resources (Jacobson et al. 2007). According to company reports of the current owner, the main pegmatite sheet at Cattlin Creek is flat lying, dipping gently to NE and is up to 21 m in thickness (Galaxy Resources Ltd. 2017). This portion represents the main mineralised part of the pegmatite (Figs. 2.13 and 2.14). Additional investigations could proof a further extension to the W, where the pegmatite sheet is covered under approximately 25–60 m of the Annabelle Volcanic rocks. Its further extension appears to be open. Continued drilling of the footwall succession up to a total depth of 232 m leads to the detection of several more, but thinner mineralised pegmatite sheets. The recent exploration activities resulted in the discovery of a SW–NE striking dolerite dike that cuts diagonally through the pegmatite sheets (Galaxy Resources Ltd. 2017). In 2010, Galaxy Lithium Australia Ltd. started mining for spodumene in an open pit mine. Total measured and indicated mineral resources are estimated to be about 10.4 Mt grading 1.23 wt% Li_2O (0.4 wt% Li_2O cut-off grade), and 151 ppm Ta_2O_5. Further 1.4 Mt are inferred resources (Galaxy Resources Ltd. 2017).

2.3 Pilbara Craton

The Pilbara Craton covers an area of approximately 60,000 km^2 in the northern part of Western Australia and is exposed in about 530 km in W–E and 230 km in N–S direction (Fig. 2.5). In contrast to the Meso- to Neo-Archean Yilgarn Craton (2880–2630 Ma; Goscombe et al. 2009), the geological record of the Pilbara Craton

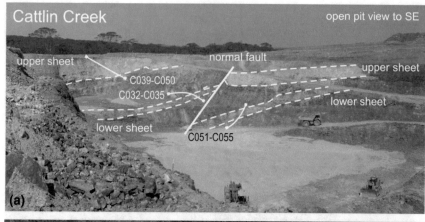

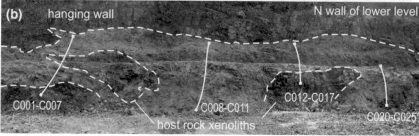

Fig. 2.14 **a** Cattlin Creek upper and lower pegmatite sheets in the open pit. View is to the SE. Numbers indicate sampling sites. **b** Sampling sites along the lower level N wall of the open pit. Two blocks of dark host rock xenoliths are displayed

represents a much older domain of Paleo- to Meso-Archean ages (3530–2830 Ma; van Kranendonk et al. 2007b). The general geological outline of the Pilbara Craton is dominated by only weakly deformed and well preserved low-grade Paleo-Archean greenstone belt successions that are interpreted to rest in classic dome-and-keel architecture between dome-like granitic complexes that were emplaced during nine distinct episodes of magmatism. Based on structures the Pilbara Craton is subdivided into three terranes, the (I) Western Pilbara Superterrane, the (II) Eastern Pilbara Terrane, and the (III) Kurrana Terrane (Fig. 2.15; Hickman 2004; van Kranendonk et al. 2007b).

The stratigraphic record of the Eastern Pilbara Craton was accumulated during two major periods from about 3530 to 3240 Ma (Pilbara Supergroup) and from 2970 to 2930 Ma (De Gray Supergroup; Smithies et al. 2005). In the Pilbara Supergroup, a cumulative stratigraphic record of up to 30 km thickness is preserved and can be subdivided into four distinct volcano-sedimentary successions. The Warrawoona Group represents the stratigraphic lowest succession, is up to 12 km thick and was deposited from about 3530 to 3430 Ma. The majority of the Warrawoona Group comprises pillow basalts, komatiitic basalt and minor komatiite. These rocks are

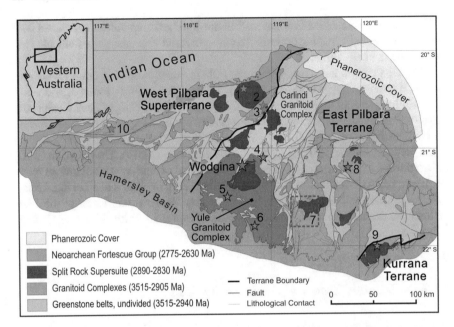

Fig. 2.15 Geological sketch map of the Pilbara Craton with the East Pilbara Terrane, West Pilbara Superterrane, Kurrana Terrane, greenstone belts, Mesoarchean Cutindua and Split Rock Supersuites, and the Wodgina pegmatite. Blue stars and broken lines indicate other pegmatite fields and deposits: 1 Pippingarra; 2 Strelley; 3 Tabba Tabba; 4 Pilgangoora; 5 Mount Francisco; 6 White Springs (Woodstock Station); 7 Shaw River; 8 Moolyella; 9 Bonney Downs; 10 Mount Hall; modified after Sweetapple and Collins (2002), Hickman (2004), van Kranendonk et al. (2006, 2007a) and Jacobson et al. (2007)

interpreted to result from several volcanogenic cycles between 3530 and 3460 Ma (van Kranendonk et al. 2007b). At about 3460 Ma a sudden change from mafic to felsic volcanism took place, which is expressed by the deposition of andesitic to rhyolitic rocks of the Panorama Formation (3458–3426 Ma). The Warrawoona Group is unconformably overlain by the Kelly Group. The stratigraphic hiatus of this erosional contact is estimated to span over a period of 75 Ma (van Kranendonk et al. 2007b). The base of the Kelly Group comprises fluviatile to shallow marine sedimentary rocks of the Strelley Pool Chert Formation. The conformably overlying Euro Basalt Formation consists of komatiite, komatiitic basalt and thoelitic basalt that were erupted from about 3350 to 3320 Ma (van Kranendonk et al. 2007b). This was followed by a short period of felsic volcanism from 3325 to 3315 Ma in the Wyman Formation (Smithies et al. 2007). The topmost section of the Kelly Group is again made up of mafic volcanic rocks of the Charteris Basalt Formation. The Sulphur Springs Group overlies the Kelly Group across an erosional unconformity. This group consists of basal sedimentary rocks of the Leilira Formation deposited between 3270 and 3250 Ma. It is conformably overlain by komatiite to komatiitic basalt of the

~3250 Ma Kunagunarrina Formation. Sulphur Spring Group volcanism closed with Kangaroo Caves Formation andesite-basalt to rhyolite volcanism at ~3240 Ma (van Kranendonk et al. 2007b).

The De Gray Supergroup overlies the Pilbara Supergroup on a craton wide erosional unconformity. Within the Eastern Pilbara Terrane this break represents a 170 Ma long stratigraphic hiatus. The stratigraphic lowest Gorge Creek Group consists of clastic sedimentary rocks, felsic tuffs and porphyries, and banded iron formation. Associated volcanism from the Western Pilbara Superterrane gives depositional ages between 3020 and 3015 Ma (van Kranendonk et al. 2007b). The Croydon Group unconformably overlies the Gorge Creek Group and comprises a basal unit of mafic volcanic rocks (Bookingarra Formation), followed by rhyolite and associated felsic volcanoclastic rocks of the Cattle Well Formation. Sedimentary rocks of the Lalla Rookh Sandstone Formation are the youngest rocks present within the Eastern Pilbara Terrane.

2.3.1 Tectonic Model of Archean Evolution

The quantity of resulting features caused by the dome-and-basin architecture is manifold. The margins of granitoid complexes commonly exhibit steeply dipping foliations parallel to, but also can pass across contacts with greenstone belts (Hickman 2004). Neighbouring greenstone belts are subjected to amphibolite- or greenschist-facies conditions in close proximity to the granitoid domes, whereas greenstone belts that are farther away from the contact exhibit still their steep dip, but were completely unaffected by contact metamorphism. Various models for the tectonic evolution of the Pilbara Craton have been proposed. One of the most recent interpretations of the development of the Pilbara Craton (Fig. A5) was compiled by van Kranendonk et al. (2007a).

D1: 3530–3460 Ma: This phase was characterised by the extrusion of large masses of basalts and komatiites that build up the predominant part of the Warrawoona Group. At about 3500 Ma, large amounts of tonalite-trondhjemite-granodiorite of the Callina Supersuite (3490–3420 Ma) intruded into the Warrawoona Group. The intrusion marks the onset of vertical tectonics that is expressed by convective overturn of the upper and middle crust and results in the juxtaposition of the Callina Supersuite granitoids contiguous to the greenstone belts. The intrusion of large amounts of granitoids is further associated with volcanism expressed by the change from mafic to felsic volcanic rocks. According to van Kranendonk et al. (2007b), the driving force for this period were multiple mantle plume events, which lead to the first phase of accumulation of large amount of continental crust within the Pilbara Craton.

D2: 3460–3420 Ma: The D2 event is interpreted to represent a second cycle of vertical tectonics during the Paleo-Archean. Again large amounts of mantle derived magma were emplaced into mid-, and upper crustal levels forming the igneous rocks of the Tambina Supersuite at 3450–3420 Ma.

D3: **3420–3300 Ma**: This phase was initiated by the partial erosion of the uppermost successions of the Warrawoona Group, expressed by a prominent unconformity at the base of the Kelly Group. Subsequent fluviatile to shallow marine sedimentary rocks (quartzite, stromatolitic marine carbonates) were deposited. At about 3350 Ma occurred the onset of another episode of mafic to ultra-mafic volcanism and the formation of thick successions of komatiitic and basaltic rocks. It was immediately followed by felsic volcanism and the emplacement of the monzogranitic Emu Pool Supersuite at 3325–3290 Ma. This supersuite is interpreted as a third cycle of vertical tectonics and crustal accumulation to the Pilbara Craton.

D4: **3270–3240 Ma**: This event is interpreted by van Kranendonk et al. (2007b) as a repetition of D3, starting with a crustal hiatus, expressed by an erosional disconformity at the base of the Sulphur Springs Group. Subsequent sedimentation of clastic and volcanoclastic material was followed by the deposition of mafic to ultramafic volcanic rocks and the onset of another cycle of crustal accumulation expressed by widespread monzogranite plutonism and related volcanism of the Clealand Supersuite rocks at 3270–3220 Ma.

D5: **3240–3050 Ma**: As the mantle plume related magmatism of D4 vanished, still upwelling hot mantle material forced the development of a continental rift and the subsequent separation of the northwestern part of the Proto-Pilbara Craton. Rift-related felsic volcanoclastic rocks, a thick succession of deep-water sedimentary rocks including clastic sedimentary rocks, pelites and banded iron formation filled the basin. This crustal extension is further associated with bimodal magmatism with emplacement of the Mount Billroth Supersuite granites at 3200–3165 Ma, and the intrusion of the mafic to ultramafic Dalton Suite. The latter encloses Ni-Cu-PGE bearing layered intrusions and associated basaltic rocks that rest in the uppermost part of the Soannesville Group. This 3200 Ma rifting event represents a major change in global tectonics. Whereas crust formation during D1–D4 was dominated by vertical tectonics related to upwelling mantle material, horizontal extension tectonics seems to become the dominant process during D5. Extension and the production of basaltic crust continued until about 3130 Ma when E–W extension was replaced by contractional plate movements towards the E. Contraction forced the development of a steep subduction zone and the production of juvenile crust within an intra-oceanic volcanic arc. Resulting melts were tonalitic in composition and now are represented by the gneisses of the Whudo Group. Furthermore, subduction likewise caused metasomatism within the overlying mantle. As subduction continued, the basin was closed completely and collision of the Eastern Pilbara Terrane with the Western Pilbara Superterrane started at about 3100 Ma. The resulting Prinsep Orogeny was accompanied by a major period of sinistral shear movement and the emplacement of the Elizabeth Hill Supersuite granites at 3090–3070 Ma (van Kranendonk et al. 2007b).

D6: **3070–2940 Ma**: This is a period of extension and concomitant deposition of the Gorge Creek Group sedimentary rocks. Extension was accompanied by crustal melting and the emplacement of the Maitland River Supersuite. Magmatism was throughout between 3020–2925 Ma and resulted in the formation of diverse rocks including boninites, high-Mg basalts, sanukitoids and alkaline granites.

D7–D9: 2970–2930 Ma: At 2940 Ma the tectonic regime changed from extension to a compressional setting and formed the ENE-trending structural pattern of the Western Pilbara Superterrane. Transpressional deformation ruled within the Eastern Pilbara Terrane. Deformation during the North Pilbara Orogeny was accompanied by the emplacement of the Sister Supersuite at 2955–2920 Ma. This magmatism was active throughout the entire Eastern Pilbara Terrane, and formed large volumes of leucogranites. According to Zegers et al. (2001) the D7–D9 deformation is further related to the formation of syntectonic pegmatites within the Eastern Pilbara Terrane.

D10: 2930–2830 Ma: As compression continued, the Kurrana Terrane was attached along the Eastern side of the Eastern Pilbara Terrane. The attachment of the Kurrana Terrane during the Mosquito Creek Orogeny was accompanied by local crustal melting, expressed by Cutinduna Supersuite intrusion at 2910–2890 Ma. Furthermore, pre-existing shear zones were reactivated within the Eastern Pilbara Terrane allowing fluid flow through the crust that locally lead to the formation of Au mineralisation (Baker et al. 2002). Of crucial importance for this study was the emplacement of the Split Rock Supersuite at 2890–2830 Ma (Fig. 2.15; van Kranendonk et al. 2007b). The Split Rock Suite Sn-Ta-Li bearing posttectonic granites intruded along a NW-trending belt that is interpreted to represent a failed rift. The Nd-model age data indicate that these granites formed due to partial melting of older granitic crust with model ages of 3400 Ma and > 3700 Ma (Smithies et al. 2003). The Sn-Ta-Li bearing granites are found across the entire Eastern Pilbara Terrane, but are significantly more abundant within the Yule, Carlindi and Pippingaara granitoid complexes and do not crop out within the central and western Pilbara Craton. Although not fully understood, the field relations and ages suggest that they are unavoidably related to the LCT pegmatites within the Eastern Pilbara Terrane. Even if only subordinate and of local importance, several generations of Proterozoic dike swarms, predominantly composed of dolerite (Wellmann 1999) and massive white monomineralic quartz veins that are locally associated with high-level epithermal Au bearing mineral systems (Huston et al. 2002) further modified the crustal framework of the Pilbara Craton. The latter is interpreted to be of Neo-Archean age (~2760 Ma).

2.3.2 The Wodgina Pegmatite District in the Pilbara Craton

The Pilbara Craton is one of the world's major tantalum provinces and contains at least 120 pegmatite deposits in over 27 pegmatite groups and fields. Mining of tantalum started in the early 20th century, principally from the Wodgina, Strelley and Tabba Tabba pegmatites. Tin is another major commodity that has been mined throughout the Pilbara Craton (e.g., Moolyella, Shaw River, Wodgina and Coondina). Although, most of the Ta and Sn was recovered from alluvial, colluvial and eluvial deposits, their primary source rare element pegmatites still have a significant economic potential. Beside Ta and Sn, also Nb, W, Be and REE have been mined as by-products (Sweetapple and Collins 2002; Jacobson et al. 2007). Spodumene and other Li-bearing minerals like lepidolite or zinnwaldite are abundant in Wodgina and

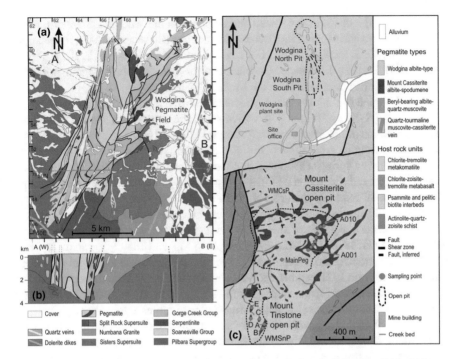

Fig. 2.16 **a** Geology of the Wodgina pegmatite deposit, modified after Hickmann (2013). **b** Cross section. **c** Geological setting of pegmatite groups at the Wodgina mining centre, modified from Sweetapple and Collins (2002)

Pilgangoora pegmatite fields. At present, no pollucite is reported. Mineral Resources Limited (MRL) is currently undertaking feasibility level engineering and design for a lithium conversion plant at Wodgina that will be capable of producing maximal 100,000 ton per annum of lithium carbonate. The first shipment of Lithium Direct Shipping Ore from Wodgina left Port Hedland towards China in April 2017.

The rare metal pegmatites in the Pilbara Craton are hosted by greenstone belt sequences that rest adjacent to granite-gneiss complexes (Fig. 2.16). Spatial distribution, as well as geochemical and geochronological data suggest that most of the rare metal pegmatites within the Pilbara Craton are related to the emplacement of the Split Rock Supersuite (2890–2830 Ma). Despite an intense exploration for rare metals during the past decades, only limited geological studies were performed on pegmatites within the Pilbara Craton (Sweetapple and Collins 2002; Jacobson et al. 2007; van Kranendonk et al. 2007a).

The Wodgina pegmatite district is located within the Pilbara Region of Western Australia at about 1200 km NNE of Perth. It consists of several dozens of individual pegmatite sheets discovered during a cassiterite mining boom from 1901 to 1905. Mining for alluvial cassiterite led to the discovery of tantalite. Since that time, the Wodgina pegmatite district was an important mining site for Sn, Ta and Be. The tenements of the district were mined and explored in the Wodgina, Stannum, Mills Find, Numbana and Mount Francisco mining centres (Sweetapple 2000; Sweetapple and Collins 2002; Jacobson et al. 2007). At the time of the field visit in 2011, Global Advanced Metals Pty Ltd. owned the tenements that belong to the Wodgina pegmatite district and mined crude ore from the Mount Cassiterite open pit.

The Wodgina pegmatite district is located within the Wodgina greenstone belt, a 25 km long (N–S) by 10 km (E–W) wide small greenstone belt surrounded by the large granitoids of the Yule Granitoid complex (Fig. 2.16a). The stratigraphic position of the Wodgina greenstone belt is still uncertain, but is considered to correlate with the Sulphur Springs (3300 Ma) and Gorge Greek Groups (3020 Ma; Hickman 1983; Blewett and Champion 2005). The contacts to granitoids of the Yule Granitoid Complex are of intrusive character at the E, whereas they are separated by a prominent sinistral shear zone at the W end of the greenstone belt. The Yule Granitoid Complex forms a melange of at least 18 distinct granitoids, subdivided into groups of pre-3400 Ma, 3270–2930 Ma, 2945 Ma, 2935 Ma, and 2850 Ma granitoids. Only the 2850 Ma granites are interpreted to have released late stage and highly fractionated melts to form LCT pegmatites (van Kranendonk et al. 2006). The sizes of the 2850 Ma granites vary from small pods (1–10 km^2) to very large bodies (100 km^2) and they were emplaced as sheets, or intruded other granites, gneisses and greenstone belts. In the Wodgina pegmatite district, this suite of granitic rocks are represented by the Gillam, Minnamonica, Numbana and Poocatche monzogranites. The largest of those granitic intrusions is the Numbana monzogranite crops out adjoining SE to the Wodgina greenstone belt. The Minnamonica and Poocatche monzogranites are situated about 15–20 km N of the Wodgina pegmatite district. The Gillam monzogranite crops out at three locations about 20 km E of the Wodgina greenstone belt.

The Wodgina mining centre includes three groups of pegmatites; the Wodgina albite type, the Mount Cassiterite albite-spodumene type, and the beryl-bearing albite-quartz-muscovite pegmatites (Fig. 2.16). The Wodgina albite type pegmatites crop out in the northern portion of the Wodgina mining centre and are hosted in metakomatiites. These pegmatites are separated by a prominent E–W striking dextral shear zone from the Mount Cassiterite albite-spodumene type pegmatites that are hosted by fine grained metapsammitic rocks. The beryl bearing albite-quartz-muscovite pegmatites in the southernmost portion of the mining centre are hosted in schists and separated from the Mount Cassiterite pegmatites by a NE–SW striking shear zone.

The Mount Cassiterite and Mount Tinstone pegmatites of the Mount Cassiterite pegmatite group (Sweetapple and Collins 2002) are located on the eastern limb of the synformal structure within the Wodgina greenstone belt (Fig. 2.16). The pegmatites are structurally constrained by a sinistral shear zone that crosscuts both the Wodgina greenstone belt, and the adjacent Numbana granite (Fig. 2.17). The two pegmatites differ by their host rocks. Metamorphosed basalts and/or ultramafic rocks host the Mount Cassiterite pegmatite, whereas the Mount Tinstone pegmatite intruded into lithologies of sedimentary origin, as metapsammites to metapelites and banded iron formations (Teitler et al. 2017). The Mount Cassiterite Group of pegmatites contains individual sheets that are 5–80 m in thickness and dip 15–25° to SW (Swee-

Fig. 2.17 **a** View to the E on the wall of the Mount Cassiterite pegmatite open pit at Wodgina. White lines indicate the locations of samples. **b** View to the SE on the Mount Cassiterite pegmatite open pit. **c** View to the W in the Mount Tinstone pegmatite open pit, with locations of sampling profiles. Blocks of host rock xenoliths are exposed

tapple and Collins 2002). Thin stringers of pegmatites interlink individual larger pegmatites sheets. The contacts to the host rocks are generally sharp, but tend to be controlled by the local stress field during emplacement, suggesting a syntectonic emplacement. Pegmatites form boudinage-like lenses that predominantly follow the general bedding of the metasedimentary rocks and are interpreted as decompressional emplacement features within a shear zone setting. The Mount Cassiterite Group pegmatites are classified as albite-spodumene-type pegmatites (Sweetapple 2000). They consist of megacrystic spodumene and perthitic microcline that rest within a matrix of medium- to fine-grained quartz, albite, K-feldspar and muscovite. Other minor minerals comprise garnet and apatite. A secondary mineral assemblage of fine-grained albite and lepidolite overprinted this mineral assemblage. Primary Ta-minerals within the pegmatite sheets are wodginite and subordinate mangano-tantalite and mangano-columbite. These minerals are further overprinted by a secondary generation of microlite and ixiolite. The cassiterite can contain up to 8 wt% Ta_2O_5 (Sweetapple and Collins 2002).

Several studies in the past tried to determine the crystallisation age of the pegmatites within the Wodgina pegmatite district. Early attempts by Jeffrey (1956) with the K-Ar- and Rb-Sr methods show a broad age range from 2890 Ma (Rb-Sr) and 2420 Ma (K-Ar) on muscovite, and 2800 Ma (Rb.Sr) and 2220 Ma (K-Ar) on microcline. More recent investigations using the U-Pb SHRIMP method by Kennedy (1998) and Kinny (2000) yielded ages of 2803 ± 115 Ma on apatite, and 2829 ± 11 Ma on tantalite, respectively. The very large range within the age data of was interpreted as open system behaviour of the isotopes due to gain or loss of parent or daughter isotopes after the primary crystallisation of the pegmatite. This process is characteristic for diffusion, or resetting of the isotope system in subsequent events. The ~2800 Ma ages suggest that the pegmatites form a part of the Split Rock Supersuite. Field relationships further support that the pegmatites represent the latest intrusion of these 2800–2850 Ma posttectonic granitoid suite. Field, geochemical, and geochronological evidence signal a genetical relationship between the Wodgina pegmatite district and the Numbana monzogranite (Sweetapple and Collins 2002; Blewett and Champion 2005).

References

Baker DEL, Seccombe PK, Collins WJ (2002) Structural history and timing of gold mineralization in the northern East Strelley Belt, Pilbara Craton, Western Australia. Econ Geol 97:775–785

Bierlein FP, Groves DI, Goldfarb RJ, Dubé B (2006) Lithospheric controls on the formation of provinces hosting giant orogenic gold deposits. Miner Depos 40:874–887

Blewett RS, Champion DC (2005) Geology of the Wodgina 1: 100,000 sheet. West Austral Geol Surv Geological Series Map 1:100,000 Explanatory Notes, 33 pp

Blewett RS, Czarnota K, Henson PA (2010) Structural-event framework for the eastern Yilgarn Craton, Western Australia, and its implications for orogenic gold. Precambr Res 182:203–229

Bucci LA, McNaughton NJ, Fletcher IR, Groves DI, Kositcin N, Stein HJ, Hagemann SG (2004) Timing and duration of high-temperature gold mineralization and spatially associated granitoid magmatism at Chalice, Yilgarn Craton, Western Australia. Econ Geol 99:1123–1144

Campell IH, Hill RI (1988) A two stage model for the formation of the granite-greenstone terrains of the Kalgoorlie-Norseman area, Western Australia. Earth Planet Sci Lett 90:11–25

Cassidy KF, Champion DC, Krapež B, Barley ME, Brown SJA, Blewett RS, Groenewald PB, Tyler IM (2006) A revised geological framework for the Yilgarn Craton, Western Australia. Geol S West Austral Record 8, 14 pp

Cassidy KF, Champion DC, McNaughton NJ, Fletcher IR, Whitaker AJ, Bastrakova IV, Budd AR (2002) Characterisation and metallogenic significance of Archaean granitoids of the Yilgarn Craton, Western Australia. Min Energy Res Inst West Austral Report 222, 536 pp

Champion DC, Cassidy KF (2007) An overview of the Yilgarn Craton and its crustal evolution. In: Bierlein FP, Knox-Robinson CM (eds) Proceedings of Geoconferences (WA) Inc: Kalgoorlie'07 Conference. Geoscience Australia Record 15:13–35

Champion DC, Sheraton JW (1997) Geochemistry and Nd isotope systematics of Archaean granites of the Eastern Goldfields, Yilgarn Craton, Australia: Implications for crustal growth processes. Precambr Res 83:109–132

Cooper DG (1964) The Geology of the Bikita pegmatite. In: Haughton SH (ed) The Geology of some ore deposits in Southern Africa, vol 1–2. Geol Soc South Africa (Johannesburg), pp 441–462

Crook D (2018) The Sinclair Zone Caesium Deposit. Pioneer Dome, WA. Austral Soc Explor Geophys. ASEG Extended Abstracts 1. https://doi.org/10.1071/aseg2018abt5_3e

Dawson GC, Krapež B, Fletcher IR, Mcnaughton NJ, Rasmussen B (2003) 1.2 Ga thermal metamorphism in the Albany-Fraser Orogen of Western Australia; consequence of collision or regional heating by dyke swarms? J Geol Soc London 160:29–37

Dirks PHGM, Jelsma HA (1998) Horizontal accretion and stabilization of the Archean Zimbabwe craton. Geology 26:11–14

Dittrich T (2016) Meso- to Neoarchean Lithium-Cesium-Tantalum- (LCT-) Pegmatites (Western Australia, Zimbabwe) and a Genetic Model for the Formation of Massive Pollucite Mineralisations. Dissertation Faculty of Geosciences, Geoengineering and Mining at TU Freiberg/Saxony Germany, 341 pp. http://nbn-resolving.de/urn:nbn:de:bsz:105-qucosa-228968

Dittrich T, Seifert T, Schulz B (2015) Genesis of selected lithium-cesium-tantalum- (LCT) pegmatites of Western Australia - with special regards to their exploration potential for the Cs-mineral pollucite and additional data from field work in the Bikita LCT pegmatite field (Zimbabwe). Final technical report, unpublished, prepared for: Rockwood Lithium GmbH, Frankfurt am Main. TU Bergakademie Freiberg, Division of Economic Geology and Petrology, 536 pp and Appendix

Dodson MH, Williams IS, Kramers JD (2001) The Mushandike Granite; further evidence for 3.4 Ga magmatism in the Zimbabwe Craton. Geol Mag 138:31–38

Dunphy JM, Fletcher IR, Cassidy KF, Champion DC (2003) Compilation of SHRIMP U-Pb geochronological data: Yilgarn Craton, Western Australia, 2001–2002. Geoscience Australia Record 15, 139 pp

Eliyahu A (2003) Controls on intrusion, crystallization and tantalum distribution of the Mount Dean rare-element pegmatite field, Norseman, Western Australia. MSc thesis University of Western Australia (unpublished)

Ellis HA (1944) A spodumene deposit, Ravensthorpe, W.A. Western Australia Department of Mines Annual Report 1943, 2 pp

Foster JG, Lambert DD, Frick LR, Maas R (1996) Re-Os isotopic evidence for genesis of Archaean nickel ores from uncontaminated komatiites. Nature 382–6593:703–706

Frei R, Blenkinsop G, Schönberg R (1999) Geochronology of the late Archaean Razi and Chilimanzi suites of granites in Zimbabwe: implications for the late Archaean tectonics of the Limpopo Belt and Zimbabwe Craton. South African J Geol 102:55–63

Galaxy Resources Ltd (2017) Galaxy Resources Ltd. Annual Report 2017. http://www.gxy.com/announcements-1/annual-report-2017

Goscombe B, Blewett RS, Czarnota K, Groenewald PB, Maas R (2009) Metamorphic evolution and integrated terrane analysis of the Eastern Yilgarn Craton: rationale, methods, outcomes and interpretation. Geol Surv West Austral Record 23, 281 pp

Griffin WL, Belousova EA, O'Neill C, O'Reilly SY, Malkovets V, Pearson NJ, Spetsius S, Wilde SA (2014) The world turns over: Hadean-Archean crust–mantle evolution. Lithos 189:2–15

Groenewald PB, Painter MGM, Roberts FI, McCabe M, Fox A (2000) East Yilgarn Geoscience Database, 1:100,000 Geology Menzies to Norseman—an explanatory note: Geol Surv West Austral Report 78, 53 pp

Groves DI, Phillips GN (1987) The genesis and tectonic controls on Archaean gold deposits of the Western Australian Shield—a metamorphic replacement model. Ore Geol Rev 2:287–322

Grubb PLC (1985) Pegmatite mineralization patterns in the Zimbabwe Craton. In: Liren W, Taiming Y, Kuirong Y, Didiers J, Grennberg JK, Lowell GR, Hongyouan X, Shoujun Y, Augustithis SS (eds) The crust—the significance of granites gneisses in the lithosphere. Theophrastus Publication S.A, Athens (Greece), pp 669–690

Gwavava O, Ranganai RT (2009) The geology and structure of the Masvingo greenstone belt and adjacent granite plutons from geophysical data, Zimbabwe Craton. South African J Geol 112:277–290

Hagemann SG, Cassidy KF (2000) Archean orogenic lode gold deposits. Rev Econ Geol 13:9–68

Hickmann AH (1983) Geology of the Pilbara Block and its environs. Geol Surv West Austral Bull 127:287

Hickmann AH (2004) Two contrasting granite-greenstone terranes in the Pilbara Craton, Australia: Evidence for vertical and horizontal tectonic regimes prior to 2900 Ma. Precambr Res 131:153–172

Hickmann AH (2013) Wodgina, WA Sheet 2655 (2nd ed). West Austral Geol Surv Geological Series Map 1: 100,000, 1 pp

Hill RI, Campbell IH, Compston W (1989) Age and origin of granitic rocks in the Kalgoorlie-Norseman region of Western Australia: implications for the origin of the Archaean crust. Geochim Cosmochim Acta 53:1259–1275

Horstwood MSA, Nesbitt RW, Noble SR, Wilson JF (1999) U-Pb zircon evidence for an extensive early Archean craton in Zimbabwe: a reassessment of the timing of craton formation, stabilization, and growth. Geology 27:707–710

Hunter WM (1988) Boorabbin, WA Sheet SH51–13 (2nd edn). West Austral Geol Surv 1:250 000 Geological Series map, 1 pp

Hunter WM (1993) Geology of the granite-greenstone terrane of the Kalgoorlie and Yilmia 1: 100,000 sheets, Western Australia. Geol Surv West Austral Report 35, 91 pp

Huston DL, Blewett RS, Keillor B, Standing J, Smithies RH, Marshall A, Mernagh TP, Kamprad J (2002) Lode gold and epithermal deposits of the Mallina basin, North Pilbara Terrain, Western Australia. Econ Geol 87:801–818

Jacobson MI, Calderwood MA, Grguric BA (2007) Guidebook to the Pegmatites of Western Australia. Hesperian Press, Carlisle/Western Australia, p 356

Jeffrey PM (1956) The radioactive age of four Western Australian pegmatites by the potassium and rubidium methods. Geochim Cosmochim Acta 10:191–195

Jelsma HA, Dirks PHGM (2002) Neoarchaean tectonic evolution of the Zimbabwe Craton. Geol Soc London Spec Publ 199:183–211. https://doi.org/10.1144/GSL.SP.2002.199.01.10

Kennedy AK (1998) SHRIMP ages of apatites from Pilbara tin-bearing pegmatites. Geol Soc Austral Abstracts 49:242

Kent AJR, Hagemann SG (1996) Constraints on the timing of lode-gold mineralization in the Wiluna greenstone belt, Yilgarn Craton, Western Australia. Austral J Earth Sci 43:573–588

Kinny PD (2000) U-Pb dating of rare metal (Sn-Ta-Li) mineralized pegmatites in Western Australia by SIMS analysis of tin and tantalum bearing ore minerals. New Frontiers in Isotope Geoscience Abstracts and Proceedings Lorne (Australia), pp 113–116

Kositcin N, Brown SJA, Barley ME, Krapež B, Cassidy KF, Champion DC (2008) SHRIMP U-Pb zircon age constraints on the Late Archaean tectonostratigraphic architecture of the Eastern Goldfields Superterrane, Yilgarn Craton, Western Australia. Precambr Res 161:5–31

Krapež B, Hand JL (2008) Late Archaean deep-marine volcaniclastic sedimentation in an arc-related basin: the Kalgoorlie Sequence of the Eastern Goldfields Superterrane, Yilgarn Craton, Western Australia. Precambr Res 161:89–113

Kusky TM (1998) Tectonic setting and terrane accretion of the Archean Zimbabwe craton. Geology 26:163–166

Martin HJ (1964) The Bikita tinfield. Southern Rhodesia Geol Surv Bull 58:114–131

McGoldrick PJ (1994) Norseman, WA Sheet 3233. West Austral Geol Surv Geological Series Map 1: 100,000, 1 pp

Melcher F, Graupner T, Gäbler HE, Sitnikova M, Henjes-Kunst F, Oberthuer T, Gerdes A, Dewaele S (2015) Tantalum-(niobium-tin) mineralisation in African pegmatites and rare metal granites: Constraints from Ta-Nb oxide mineralogy, geochemistry and U-Pb geochronology. Ore Geol Rev 64:667–719. https://doi.org/10.1016/j.oregeorev.2013.09.003

Mikucki EJ, Roberts FI (2004) Metamorphic petrography of the Kalgoorlie region, Eastern Goldfields Granite-Greenstone Terrane: METPET database. Geol Surv West Austral Record 12, 40 pp

Myers JS (1993) Precambrian history of the West Australian Craton and adjacent orogens. Ann Rev Earth Planet Sci 21:453–485

Nelson DR (1995) Compilation of SHRIMP U-Pb geochronology data, 1994. Geol Surv West Austral Record 3, 251 pp

Nelson DR (1997) Evolution of the Archaean granite-greenstone terranes of the Eastern Goldfields, Western Australia: SHRIMP U-Pb zircon constraints. Precambr Res 83:57–81

Oberthuer T, Davis DW, Blenkinson TG, Höhndorf A (2002) Precise U-Pb mineral ages, Rb-Sr and Sm-Nd systematics for the Great Dyke, Zimbabwe—constraints on late Archean events in the Zimbabwe Craton. Precambr Res 113:293–305

Partington GA, McNaughton NJ, Williams IS (1995) A review of the geology, mineralization, and geochronology of the Greenbushes Pegmatite, Western Australia. Econ Geol 90:616–635

Pehrsson SJ, Berman RG, Eglington B, Rainbird R (2013) Two Neoarchean supercontinents revisited: the case for a Rae family of cratons. Precambr Res 232:27–43

Poineer Resources Ltd (2017) Mineral resource estimate for the Sinclair Cesium Project. http://www.pioneerresources.com.au/project_pioneerdome.php

Prendergast MD (2004) The Bulawayan Supergroup: a late Archaean passive margin-related large igneous province in the Zimbabwe Craton. J Geol Soc London 161:431–445

Prendergast MD, Wingate MTD (2007) Zircon geochronology and partial structural re-interpretation of the late Archaean Mashaba Igneous Complex, south-central Zimbabwe. South African J Geol 110:585–596

Roddick JC (1974) Responses of strontium isotopes to some crustal processes. Ph.D. thesis Australian National University Canberra, 350 pp (unpublished)

Rollinson HR, Whitehouse M (2011) The growth of the Zimbabwe Craton during the late Archaean: an ion microprobe U-Pb zircon study. J Geol Soc London 168:941–952

Savage M, Barley ME, McNaughton NJ (1995) SHRIMP U-Pb geochronology of 2.95–3.0 Ga felsic igneous rocks at Ravensthorpe, Yellowdine Terrane, Yilgarn Craton. Australian Conference on Geochronology and Isotope Geoscience, Workshop Programme and Abstracts. Curtin University of Technology Perth, 30 pp

Scibiorskia E, Tohver E, Jourdan F (2015) Rapid cooling and exhumation in the western part of the Mesoproterozoic Albany-Fraser Orogen, Western Australia. Precambr Res 265:232–248

Shimizu K, Nakamura E, Maruyama S (2005) The geochemistry of ultramafic to mafic volcanics from the Belingwe greenstone belt, Zimbabwe: Magmatism in an Archean continental large igneous province. J Petrol 46:2367–2394

Smithies RH, Champion DC, Cassidy KF (2003) Formation of Earth's early Archaean continental crust. Precambr Res 127(1–3):89–101

Smithies RH, Champion DC, van Kranendonk MJ, Hickman AH (2007) Geochemistry of volcanic rocks of the northern Pilbara Craton, Western Australia. Geol Surv West Austral Report 104, 47 pp

Smithies RH, Champion DC, van Kranendonk M, Howard HM, Hickman AH (2005) Modern-style subduction processes in the Mesoarchaean; geochemical evidence from the 3.12 Ga Whundo intra-oceanic arc. Earth Planet Sci Lett 231:221–237

Smithies RH, Ivanic TJ, Lowrey JR, Morris PA, Barnes J, Wyche S, Lu YJ (2018) Two distinct origins for Archean greenstone belts. Earth Planet Sci Lett 487:106–116

Smithies RH, Spaggiari CV, Kirkland CL (2015) Building the crust of the Albany-Fraser Orogen: constraints from granite geochemistry. Geol Surv West Austral Report 150, 49 pp

Soederlund U, Hofmann A, Klausen MB, Olsson JR, Ernst RE, Persson PO (2010) Towards a complete magmatic barcode for the Zimbabwe craton: baddeleyite U-Pb dating of regional dolerite dyke swarms and sill complexes. Precambr Res 183:388–398

Sofoulis J (1958) Report on Cattlin Creek spodumene pegmatite, Ravensthorpe, Phillips River Goldfield. Geol Surv West Austral Bull 110:193–202

Stark JC, Simon A, Wilde SA, Söderlund U, Li ZX, Rasmussen B, Zi JW (2018) First evidence of Archean mafic dykes at 2.62 Ga in the Yilgarn Craton, Western Australia: Links to cratonisation and the Zimbabwe Craton. Precambr Res 317:1–13

Sweetapple MT (2000) Characteristics of Sn-Ta-Be-Li-industrial mineral deposits of the Archaean Pilbara Craton, Western Australia. Austral Geol Surv Org Record 2000–44, 59 pp

Sweetapple MT (2017) A review of the setting and internal characteristics of lithium pegmatite systems of the Archaean North Pilbara and Yilgarn Cratons, Western Australia. Granites 2017 Conference Benalla Victoria, Ext Abstr Austr Inst Geosci Bull 65

Sweetapple MT, Collins PLF (2002) Genetic framework for the classification and distribution of Archean rare metal pegmatites in the North Pilbara Craton, Western Australia. Econ Geol 97:873–895

Taylor PN, Kramers JD, Moorbath S, Wilson JF, Orpen JL, Martin A (1991) Pb/Pb, Sm-Nd and Rb-Sr geochronology in the Archean craton of Zimbabwe. Chem Geol 87:175–196

Teitler Y, Duuring P, Hagemann SG (2017) Genesis history of iron ore from Mesoarchean BIF at the Wodgina mine, Western Australia. Austral J Earth Sci 64:41–62. https://doi.org/10.1080/08120099.2017.1266387

Thom R, Lipple SL (1977) Ravensthorpe, WA Sheet SI51-5. Geol Surv West Austral 1: 250 000 Geological Series map, 1 pp

Turek A (1966) Rb-Sr isotopic studies in the Kalgoorlie-Norseman area, Western Australia. Ph.D. thesis Australian National University (unpublished)

van Kranendonk MJ, Hickman AH, Smithies RH, Williams IR, Bagas L, Farrell TR (2006) Revised lithostratigraphy of Archean supracrustal and intrusive rocks in the northern Pilbara Craton, Western Australia. Geol Surv West Austral Record 15, 63 pp

van Kranendonk MJ, Smithies RH, Hickman AH, Champion DC (2007a) Paleoarchean development of a continental nucleus: The East Pilbara Terrane of the Pilbara Craton, Western Australia. In: van Kranendonk HJ, Smithies RJ, Bennett VC (eds) Earth's Oldest Rocks. Developments in Precambrian Geology vol 15, Elsevier Amsterdam, pp 307–337

van Kranendonk MJ, Smithies RH, Hickman AH, Champion DC (2007b) Review: secular tectonic evolution of Archean continental crust: interplay between horizontal and vertical processes in the formation of the Pilbara Craton, Australia. Terra Nova 19:1–38

Wellmann P (1999) Interpretation of regional geophysics of the Pilbara Craton, northwest Australia. Austral Geol Surv Org Record 4, 60 pp

Wilson JF (1964) The geology of the country around Fort Victoria. Southern Rhodesia Geol Surv Bull 58, 147 pp

Wilson JF (1990) A craton and its cracks; some of the behaviour of the Zimbabwe Block from the late Archaean to the Mesozoic in response to horizontal movements, and the significance of some of its mafic dyke fracture patterns. J African Earth Sci 10:483–501

Wilson JE, Nesbitt RW, Fanning CM (1995) Zircon geochronology of Archaean felsic sequences in the Zimbabwe Craton: a revision of greenstone stratigraphy and a model for crustal growth. In: Coward MP, Ries AC (eds) Early Precambrian Processes. Geol Soc London Spec Publ 95:109–126

Wilson JF, Orpen JL, Bickle MJ, Hawkesworth CJ, Martin A, Nisbet EG (1978) Granite-greenstone terrains of the Rhodesian Archaean craton. Nature 271–5640:23–27

Witt WK (1996) Ravensthorpe, WA Sheet 2930. Geol Surv West Austral Geol Series 1: 100,000 map, 1 pp

Witt WK (1999) The Archaean Ravensthorpe Terrane, Western Australia: synvolcanic Cu-Au mineralization in a deformed island arc complex. Precambr Res 96:143–181

Witt WK, Cassidy KF, Lu Y-J, Hagemann SG (2017) Syenitic Group intrusions of the Archean Kurnalpi terrane, Yilgarn craton: Hosts to ancient alkali porphyry gold deposits? Ore Geol Rev. https://doi.org/10.1016/j.oregeorev.2017.08.037

Witt WK, Cassidy KF, Lu Y-L, Hagemann SG (2018) The tectonic setting and evolution of the 2.7 Ga Kalgoorlie-Kurnalpi Rift, a world-class Archean gold province. Mineral Depos. https://doi.org/10.1007/s00126-017-0778-9

Wyche S, Kirkland CL, Riganti A, Pawley MJ, Belousova E, Wingate MTD (2012) Isotopic constraints on stratigraphy in the central and eastern Yilgarn Craton, Western Australia. Austral J Earth Sci 59:657–670

Zegers TE, Nelson DR, Wijbrans JR, White SH (2001) SHRIMP U-Pb zircon dating of Archean core complex formation and pancratonic strike-slip deformation in the East Pilbara granite-greenstone terrain. Tectonics 20:883–908

Chapter 3
Petrography and Mineralogy

For petrography, mineralogy and microstructural characterisation of pegmatites, transmitted and reflected light microscopy was conducted on 307 polished thin sections. The observations were precised by backscattered electron imageing (BSE) and energy dispersive X-ray spectroscopy (EDS) analyses with a scanning electron microscope SEM at Geometallurgy Laboratory of the Department of Economic Geology and Petrology at TU Freiberg in Saxony/Germany. The instrument was a FEI Quanta 600-FEG-MLA, equipped with two Bruker X-Flash SDD-EDS X-ray spectrometers. A Bruker Software ESPRIT Quantax 200 was used for processing of EDS-data and calculation of the chemical compositions. As pegmatites are coarse-grained rocks, the modal mineralogy as obtained from thin sections (25 × 45 mm) is not representative for single samples. For the whole rock bulk geochemical analyses, larger rock plates of 2–3 kg were comminuted and then homogenised. One part was powdered for geochemical analyses. An unpowdered aliquot with particles up to 200 μm in diameter was embedded in epoxy blocks (30 mm diameter) and polished. Then SEM-based automated mineralogical methods of the FEI-MLA 2.9 mineral liberation analysis software version were applied for modal analysis. A set of reference spectra from standards and from pegmatite minerals was used for classification of the measurements. Main advantage of the mineral liberation analysis with respect to alternative methods as XRD (X-ray diffraction) is its capability to identify accessory minerals at <1 wt% of bulk modal composition, as cassiterite, monazite, zircon and Nb–Ta oxides within a sample. Also the SEM-MLA technique allows to demonstrate the presence of Cs-bearing rare minerals, as Cs-enriched lepidolite, nanpingite, sokolovaite, pezzottaite, and even small grains of dispersed pollucite. Elements as Li and Be cannot be detected by electron beam based analytics. This hampers the identification of Li- and Be-minerals, such as petalite, spodumene or beryl. Careful inspection of Al–Si ratios in the EDS-spectra in combination with elemental analysis and optical microscopy allowed recognition of these minerals. A total of 81 different mineral species were identified and summarised (Table A29). Based on the modal mineralogy by mineral liberation analysis (MLA) methods, the samples were classified into 21 distinct mineral associations (Fig. 3.1). Monominer-

T. Dittrich et al., *Archean Rare-Metal Pegmatites in Zimbabwe and Western Australia*,
SpringerBriefs in World Mineral Deposits, https://doi.org/10.1007/978-3-030-10943-1_3

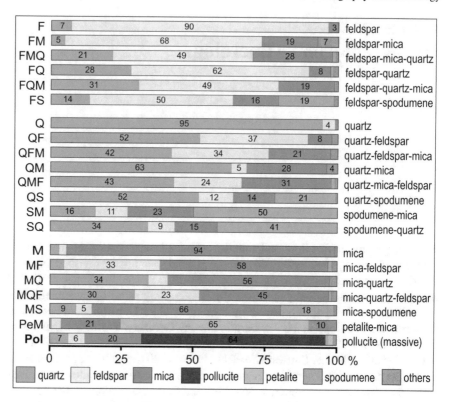

Fig. 3.1 Compositions of the 21 mineral association types, as quantified by automated SEM-Mineral Liberation Analysis (MLA) methods. Abbreviations to the left correspond to mineral associations to the right

alic mineral associations (F, M, Q) contain at least 80 wt% of a mineral. The other mineral associations are named according to the minerals that made up at least 80 wt% of the sample. An exception is pollucite where all samples with at least 15 wt% pollucite are classified as pollucite mineral assemblage. For the other mineral associations only minerals with at least 15 wt% were considered.

3.1 Minerals in LCT Pegmatites

3.1.1 Feldspars

Feldspars are predominant in the majority of the pegmatite samples. Plagioclase is the most common feldspar. Albeit K-feldspars such as orthoclase or microcline are only a subordinate to minor component of the LCT pegmatites, they can be of local importance. The plagioclases are mostly albite (Na), with anorthite (Ca) being almost completely absent. Albite occurs in monomineralic masses, single crystals up to 30~cm long or the fine-grained matrix between quartz, mica and spodumene. Fine-grained varieties display saccharoidal or aplitic textures. In most of the samples, three generations of albite are present. The first generation is characterised by almost euhedral to subhedral fine-grained to megacrystic predominantly inequidimensional crystals with typical albite twinning (Fig. A6). They are intergrown with anhedral to subhedral muscovite and lepidolite and anhedral quartz. Within the Cattlin Creek pegmatite and the Wodgina pegmatite, albite furthermore is observed in a preferred orientation that is interpreted to represent magmatic lamination. The second generation of albite are xenomorphic grains interstitial to the first generation albites, as well as interstitial to quartz, feldspar and spodumene. The third generation of albite replaces orthoclase, microcline and mica. When albite replaces K-feldspar it is locally present as pale blue cleavelandite that is characteristically associated to lepidolite. This may be interpreted as a late stage magmatic or hydrothermal overprint. Such late stage processes can be also replace albite by white mica, as lepidolite, muscovite or sericite. Replacement of albite by mica is initiated along cleavage planes and fractures and is accompanied by crystallisation of quartz (Dittrich 2016).

K-feldspar is the major component in samples BQ09, C047b, L009, L018, MD119, WMMainA, WMSnP-B006 and subordinate to minor components within most of the other samples. As a major component, K-feldspar occurs in white to grey agglomerations composed of hypidiomorphic to xenomorphic grains intergrown with mica, quartz, albite, and spodumene. Microcline exhibits the typical twinning pattern and is often replaced by albite, in part cleavelandite, lepidolite and also fluorite, especially at Wodgina. The EDS-analyses reveal some enrichment of BaO and Rb_2O for several relic crystals of K-feldspar in samples L015, L073, C011 which is interpreted to represent a late (hydrothermal) replacement stage (Fig. A7).

The petrography of the feldspars allows some conclusions concerning their crystallisation in a pegmatitic melt: Almost euhedral to subhedral crystal shapes of the first plagioclase generation indicate that they represent the beginning of the crystallisation of the melt. The anhedral to xenomorphic grain shape of plagioclase from the second generation is interpreted to have been formed by partial resorption of the first generation plagioclase by changing physicochemical conditions within the melt. These changes may be induced by ascend of the pegmatite magma, which can also explain the magmatic lamination observed in various samples. Other potential processes are release of fluid or addition of new magma. The third generation of albite is interpreted to be related to late stage hydrothermal processes after the consolidation of the pegmatite. As for the plagioclase, the structures of the K-feldspar suggest that it crystallises relatively early from the pegmatite melt and was later affected by resorption due to changing physicochemical conditions. This early crystallisation further supports that a fluid release from the pegmatite magma could explain the resorption observed in the first generation of plagioclase and the K-feldspar. The feldspars are replaced by micas during a late stage. This is interpreted to result either by subsolidus self-ordering processes or by a deuteric or hydrothermal alteration.

3.1.2 Quartz

In the LCT pegmatites, quartz occurs as clear, white, and with partly milky to grey and smoky grains. Grain sizes range from tiny submicrometer inclusions up to larger polycrystalline aggregates of several dozens of centimeters in diameter. Quartz forms mostly anhedral grains that rest in interstitial positions between euhedral to subhedral albite, K-feldspar, muscovite and lepidolite. This textural position suggests that quartz postdates the crystallisation of these minerals. Quartz is also observed to replace feldspars and mica. Larger polycrystalline quartz aggregates are enclosed within a matrix of albite, K-feldspar and mica. Another characteristic texture is the intergrowth of quartz and mica represented by the QM- and MQ-mineral assemblages where both minerals form equigranular bands close to the contact zone or form circular masses that rest within huge masses of monomineralic feldspar or feldspar-quartz intergrowth (Fig. A8). Quartz is also observed to form symplectitic intergrowth with albite or mica that appears as seam around larger grains of feldspar, quartz or spodumene. Fluid inclusions and inclusions of all other minerals, enclosing the accessories are abundant in quartz. These textural relationships are interpreted to result from the primary crystallisation of quartz in the melt. There are also indications for a late stage and post-magmatic formation of quartz in a great number of samples. Late-stage quartz occurs in small veins crosscutting earlier formed minerals like feldspars, pollucite and spodumene or is found associated with mica replacing albite and K-feldspar along small fissures or within cleavage plains. Another post-magmatic microstructural feature in a large number of samples are recrystallisation textures such as undulose extinction, subgrain migration and briquette-like crystal shapes that are rather typical for deformation and hydrothermal crystallisation.

3.1.3 Mica

Mica is present in almost all samples and occurs as muscovite and lepidolite, but also in other species such as zinnwaldite, trilithionite or bityite. Furthermore, micas are present in at least two to three generations, (early) magmatic, late magmatic to hydrothermal and late stage alteration and sericitisation. **Muscovite** occurs in large pale to dark green coloured idiomorphic crystals up to several centimeters in diameter. It typically forms single crystals or flakes intergrown with quartz or feldspar and is abundant in the contact or border zones of the pegmatites. Smaller crystals (<5~mm) were observed in the central portions of the pegmatite. There, muscovite is intergrown with almost all other minerals such as quartz, feldspar or spodumene (Fig. A9). In part, muscovite exhibits symplectitic intergrowth with quartz or feldspar. This textural feature is especially observed at grain boundaries of large quartz, feldspar or spodumene crystals, respectively. SEM inspection indicates that the rim of certain muscovite is enriched in up to several wt% Cs in a large number of samples. This signals enrichment of Cs during late stage magmatic to hydrothermal processes. The textural positions of muscovite are indicative for late stage replacement processes. This generation of muscovite occurs along veins or in cleavage plains of feldspar and spodumene or within cleavage plains of earlier formed muscovite. At the Londonderry and Wodgina pegmatites, muscovite is further accompanied by fluorite that rests within the cleavage plains. **Lepidolite** appears to be restricted to the inner or higher evolved portions of the pegmatites. At Bikita, lepidolite is present in larger monomineralic masses. The colours of lepidolite exhibit a large spectrum ranging from dark purple to magenta, bright pink to clear. Lepidolite is mostly present as fine to medium grained, rarely >1 cm crystals. It occurs as tabular anhedral to irregular subhedral shaped grains that are intergrown with almost all observed minerals. In addition, lepidolite is observed as book-like, radial or nest-like aggregates. Lepidolite can replace albite and K-feldspar along grain boundaries, cleavage plains or fractures. When it replaces larger K-feldspar crystals or aggregates, it is associated by blue cleavelandite. Another characteristic feature of lepidolite are fracture fillings within the massive pollucite mineralisation at Bikita. Within these veins, lepidolite is associated with petalite, albite and quartz. A similar vein filling is observed in certain feldspars, petalite and spodumene at the Londonderry pegmatite (Lepidolite Hill). As for muscovite, the SEM revealed enrichment in Cs (BQ22, 1–8 wt% Cs) for some lepidolite rims. At Bikita, some lepidolites exhibit lamellae enriched up to 20 wt% of Cs (Fig. A10). The structural position of lepidolite within the inner or higher evolved portions of the pegmatite suggests that the formation of lepidolite is restricted to late stage magmatic or hydrothermal processes.

3.1.4 Pollucite

Pollucite was observed at the Bikita Main Quarry and the Dam Site (Fig. 2.4a, d). At both locations it forms lens-like massive mineralisation. Despite the occurrence of some small crystals within a matrix of feldspar and mica, pollucite exclusively occurs in massive monomineralic masses. Pollucite is white to slight grey coloured and characterised by a network of submillimeter to centimeter wide veins composed of purple lepidolite, petalite, quartz and feldspar. Furthermore, pollucite exhibits a broad spectrum of mineral inclusions such as quartz, petalite, albite (in part cleavelandite) and K-feldspar. Individual inclusions reach up to several dozen of centimeters in diameter. The massive pollucite mineralisation rests in sharp contact towards surrounding feldspar units and mineral inclusions. In the case of petalite, pollucite is present as symplectitic intergrowth with petalite in small cracks that crosscuts the petalite predominantly at its cleavage plains. Combined BSE and EDS-analysis by SEM demonstrate that pollucite exhibits a rather inhomogeneous composition. The Cs contents range from about <20 wt% Cs_2O up to 37 wt% Cs_2O. Under BSE-imaging, this variation is expressed by different grey values, where brighter regions represent zones of higher Cs concentration. The zonation is predominantly patchy, but higher Cs concentrations are preferentially found along the lepidolite-quartz-petalite veins that crosscut the massive lepidolite, along boundaries to various mineral inclusions and along several subordinate generations of cracks (Fig. 3.2). Thus, these smaller scaled cracks and networks of lepidolite-quartz-petalite veins are interpreted as fluid pathways and evident for a late stage Cs-enrichment. Another characteristic feature of the massive pollucite samples is that these are almost completely lack Nb–Ta–Sn-oxide minerals. The only mineral of this group that was identified in all massive pollucite samples are small traces of Sb-bearing tantalite. Even though pollucite is known from the Londonderry pegmatite field, it was not found during the two field campaigns. A small specimen was provided by the Western Australian Mineral Collection from the original diamond drill core. EDS-analysis confirmed pollucite. In sample L062 tiny inclusions of mineral phases within quartz were detected that contain up to 30 wt% Cs. These Cs rich inclusions are too small in order to allow a safe classification of pollucite.

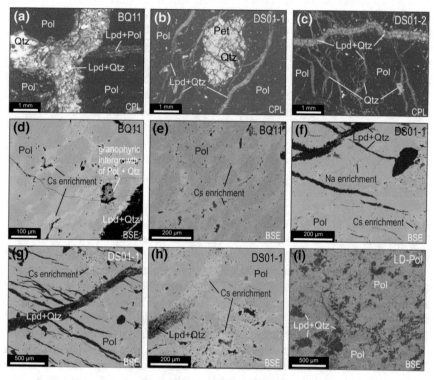

Fig. 3.2 Microstructures of pollucite (Pol) in Bikita pegmatite, in cross polarized light (CPL) and backscattered electron (BSE) images. **a** Pollucite is crosscut by veins of lepidolite (Lpd) and quartz (Qtz). **b** Pollucite with inclusion of petalite (Pet) and quartz, and a network of lepidolite and quartz. **c** Pollucite is crosscut by several generations of veins of lepidolite or quartz, and a second generation of pollucite. **d** Pollucite with brighter areas of Cs enrichment. Granophyric intergrowth of pollucite and quartz. **e** Pollucite with brighter areas representing several stages of Cs enrichment. **f, g** Pollucite crosscut by several generations of veins with lepidolite and quartz. Brighter areas indicate Cs enrichment. **i** Pollucite, displaying patchy aggregates composed of lepidolite and quartz

3.1.5 Petalite

The Li-mineral petalite was only encountered within the Bikita and Londonderry pegmatite fields. It typically occurs in white subhedral to almost euhedral crystals that rest within albitic units of the pegmatite. Although very similar to the surrounding albite, petalite can be easily distinguished by its prominent (001) cleavage plane. Within the Londonderry Feldspar Quarry pegmatite, petalite is also observed as white to slight pastel-rose altered mineral masses. Petalite is further characterised by tiny inclusions of lepidolite, albite and quartz. At Bikita fractures within petalite are filled by a symplectitic intergrowth of petalite and pollucite. In addition, huge subhedral crystals of petalite are also found within the massive pollucite mineralisation of the Bikita Main Quarry.The subhedral to almost euhedral shape of the petalite and the structural position as single crystals or masses within the albitic units suggest that the petalite was formed prior or contemporaneously to the feldspars during the initial stage of crystallisation (Fig. 3.3). A subsolvus crystallisation under elevate water pressures can be suggested from the presence of both feldspars in the early crystallisation stage.

3.1.6 Spodumene

The Li-mineral spodumene was encountered at Cattlin Creek, Wodgina, and Bikita. It is found in the intermediate and core zones of the Cattlin Creek pegmatite, in the core zone of the Wodgina pegmatite and at the Bikita pegmatite field within the central zone. Spodumene occurs in pale green (Cattlin Creek) to yellowish-brown (Wodgina) to white (Bikita) colours. At Cattlin Creek, some of the spodumene exhibits purple colour. It characteristically forms euhedral crystals up to several dozen of centimeters, and for the case of the Cattlin Creek even meters in size that rest within a matrix of predominantly albite and subordinate quartz and mica. The euhedral grains contain inclusions of albite, in part cleavelandite, tourmaline, quartz, muscovite and lepidolite, as well as numerous fluid inclusions (Fig. 3.4). According to textural and structural relationships, the mineral inclusions belong to two generations: Albite and quartz are interpreted as primary inclusions that were entrapped during the crystal growth of spodumene. Inclusions of tourmaline, lepidolite, cleavelandite and muscovite replace preexisting spodumene. As observed for the Li mineral petalite, the structural positions and the subhedral to almost euhedral shape of the spodumene suggest that it formed prior or contemporaneous to the feldspars during the initial crystallisation stage of the melt. Primary magmatic pyroxenes are interpreted to result from hydrous melts (France et al. 2013) which is in good agreement with the suggested subsolvus crystallisation under elevated water pressures. An observed resorption of spodumene by other minerals like mica, albite or quartz can be interpreted by a change in physicochemical conditions in a later stage of crystallisation (Novák et al. 2015). This also can lead to the formation of the so called spodumene quartz intergrowth (Thomas et al. 1994).

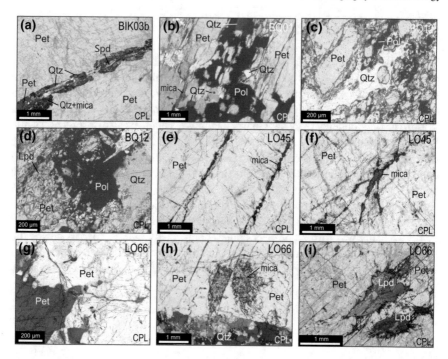

Fig. 3.3 Microstructures of petalite (Pet) in polarised (PL) and cross polarised light (CPL). **a** Petalite crossed by a prominent vein of quartz (Qtz), mica and spodumene (Spd). **b** Petalite partly replaced by pollucite (Pol) along cleavage planes or small fractures. **c** Petalite is replaced by quartz and pollucite (Pol) along cleavage plains or small fractures. **d** Petalite intergrown with pollucite quartz and lepidolite (Lpd). **e** Petalite is crosscut by small veins filled with mica; **f** Petalite is crossed by several sets of veins and replaced by a mica aggregate. **g** Aggregate of petalite is crossed by several generations of veins. **h** Quartz in contact to and replacing petalite. **i** Petalite contains several aggregates and veins of lepidolite. Replacement of petalite along the cleavage planes

3.1.7 Beryl

Specimens of Beryl were observed at the Cattlin Creek (sample C013), Londonderry (at LPD-Hill Brl 1) and Wodgina. Within the Cattlin Creek pegmatite, beryl occurs as minor phase that forms pale greenish to yellow coloured hypidiomorphic crystals up to several centimeters in diameter. It rests in interstitial position between albite, quartz, muscovite, lepidolite and minor spodumene (Fig. 3.5). The SEM study revealed three generations of beryl in the pegmatites: The first generation of beryl crystallised with the surrounding albite and quartz, as indicated by inclusions of albite and quartz. The second generation of beryl exhibits an enrichment of up to 1 wt% Cs_2O. This beryl crystallises as rim around and in fissures crosscutting the beryl 1. Furthermore, beryl encloses secondary Be-minerals as bertrandite, bavenite and bityite. These occur along fissures and exhibit a cloudy appearance. They are associated to a third generation of beryl with an extreme enrichment of up to 10–15 wt%

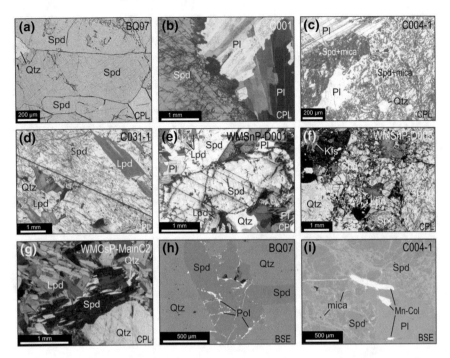

Fig. 3.4 Microstructures of spodumene (Spd) in cross polarised light (CPL) and backscattered electron images (BSE). **a** Aggregate of hypidiomorphic spodumene intergrown with minor amounts of quartz (Qtz). Prominent cleavage plains of spodumene. **b** Spodumene in sutured contact to plagioclase (Pl). **c** Spodumene replaced by fine grained mica. **d** Spodumene with inclusions of lepidolite (Lpd) and quartz. **e** Hypidiomorphic spodumene intergrown with lepidolite, quartz and plagioclase. **f** Spodumene replaced by lepidolite and quartz. **g** Xenomorphic remnants of spodumene in interstitial position between lepidolite and quartz. **h** Spodumene and quartz intergrowth. Note the small aggregates of pollucite (Pol) within cracks of the spodumene. **i** Spodumene that is replaced by fine grained mica

Cs_2O. Within the Londonderry pegmatite, beryl occurs as fractured crystals up to 5 mm in diameter that rest within a matrix composed of albite, muscovite and quartz. The latter are also enclosed in beryl. The Cs-enrichment of up to 1.7 wt% Cs_2O is only detectable in some minor domains at the rim of the beryl crystals or along small fissures. As for the Cattlin Creek pegmatite, this enrichment is interpreted to represent a second generation of beryl growth. The position in interstices suggests that beryl 1 crystallised after most of the plagioclase and K-feldspar as well as spodumene within the Cattlin Creek pegmatite. The formation of the second generation of beryl with elevated Cs requires the presence of a Be–Cs enriched fluid. This is in agreement with the subsolvus crystallisation under elevated water pressures and a subsequent change in physicochemical conditions of the melt.

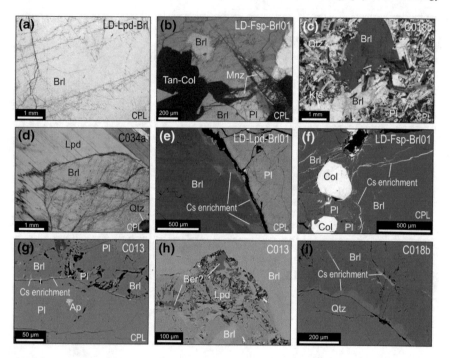

Fig. 3.5 Microstructures of beryl (Brl) in polarised (PL) and cross polarised light (CPL), and in backscattered electron imageing (BSE). **a** Fractured beryl. **b** Beryl with inclusions of plagioclase, tantalite-columbite (Tan-Col) group minerals and idiomorphic monazite (Mnz). **c** Beryl intergrown with plagioclase (Pl), quartz (Qtz) and K-feldspar (Kfs). **d** Fractured beryl in contact to lepidolite (Lpd) and quartz; **e** Beryl and plagioclase; zones of Cs enrichment along the margin of beryl. **f** Beryl with inclusions of columbite (Col) and plagioclase. Enrichment of Cs along small fissures. **g** Beryl intergrown with feldspar, quartz and bertrandite (Ber). Aggregates of lepidolite and bertrandite within beryl. **i** Beryl and quartz with enrichment of Cs along the contact and fractures

3.1.8 Tourmaline

Tourmaline was only observed at the Cattlin Creek pegmatite were it is present in all zones and almost all mineral assemblages (Fig. A11). It exhibits a bright colour range from black to bluish and pale-purple with a characteristically zonation of black cores and blue rims. Tourmaline is observed as euhedral grains within sugary albite but predominantly rests in interstitial position where it is intergrown with albite, K-feldspar, muscovite and spodumene. Tourmaline is observed in skeletal or poikiloblastic grains that contain abundant mineral inclusions of plagioclase and quartz. The EDS analyses mostly indicate a schorl composition. Data obtained from bluish rims reflect elbaitic to dravitic or olenitic composition. Most of the tourmaline rims are characterised by small amounts of Cs. The presence of tourmaline throughout the whole pegmatite sheet indicates that it could crystallise through the complete crystallisation sequence. However, its textural position in subhedral grains suggest a crystallisation subsequent to plagioclase and K-feldspar. This temporal range is further supported by the replacement of spodumene by tourmaline, quartz and muscovite.

3.1.9 Apatite

Apatite is present as minor to accessory component in a large variety of samples. It occurs as dark grey to greenish to turquoise blue crystals or aggregates up to 5 mm in size (Fig. A12). Apatite occurs in three structural settings. Inclusions of apatite in massive K-feldspar, albite and quartz testify a primary crystallisation from the pegmatitic melt. The second structural positions where apatite is found in distinct grains and aggregates between albite or in interstitial position between feldspars and mica, indicates replacement at a late stage magmatic or hydrothermal processes. In the third structural setting apatite occurs as filling of small fractures and within cleavage plains of late stage micas. The structural position of apatite as inclusion within or in interstitial position between K-feldspar, plagioclase and quartz suggests that it was formed contemporaneous or short after the crystallisation of the host or associated minerals within the early stages of magmatic crystallisation. In contrast, the third structural setting as filling of veins or fractures suggests a late stage formation most probably from a fluid.

3.1.10 Ta-, Nb- and Sn-Oxides

Most common are tantalite, columbite and cassiterite. Other Ta-, Nb- and Sn-oxide minerals ascertained during the study are microlite, wodginite, ixiolite and pyrochlor. They occur in various textural and structural positions that are assigned to at least two distinct periods of growth: (1) primary crystallisation from a melt or fluid (tantalite, columbite and cassiterite); (2) hydrothermal replacement of the primary Ta-, Nb- and Sn-oxides by microlite, ixiolite and wodginite. The bulk of the cassiterite occurs as hypidiomorphic to idiomorphic grains up to several millimeters in diameter, predominantly as inclusions within albite, K-feldspar, quartz and spodumene and is frequently is accompanied by columbite and tantalite or their secondary replacement products (Fig. A13). Some grains exhibit compositional zoning under BSE imageing. Cassiterite contains up to 2 wt% Ta.

Columbite and tantalite predominantly are observed as euhedral to subhedral crystals with sizes ranging between a few micrometer up to several millimeter in diameter (Fig. A14). They form inclusions within feldspar and are positioned between quartz, feldspar or mica grains. Columbite and tantalite also rest within the cleavage plains of certain micas. The chemical compositions range from ferrocolumbite and ferrotantalite to mangano-columbite and mangano-tantalite, respectively. The chemical variability reflects different stages of magmatic (hydrothermal) fractionation. In addition to Mn and Fe, varieties of Bi and Sb bearing tantalite were observed at the Cattlin Creek pegmatite, as well as within the Londonderry and Bikita pegmatite fields. Furthermore, columbite and tantalite are associated with cassiterite and secondary Nb–Ta oxide minerals like microlite or ixiolite (Fig. A15). The secondary Ta-, Nb-, and Sn-oxides are present as microlite, wodginite, ixiolite or pyrochlor. They form irregular shaped grains or complex aggregates that overprint and replace primary tantalite and columbite. Replacement of the primary assemblages is visible as coatings, by growth of secondary minerals along fractures, or by precipitation interstitial to feldspar, quartz and mica. Resorption of the tantalite and columbite is likewise related to the replacement of the primary Nb–Ta–Sn mineral assemblage by microlite, wodginite, ixiolite or pyrochlor (Dittrich 2016).

3.2 Reconstruction of the General Crystallisation Sequence

Three distinct crystallisation stages can be recognised in the LCT pegmatites (Fig. 3.6). The early stage is characterised by minerals with predominantly euhedral crystal shapes, as plagioclase and K-feldspar, as well as petalite, spodumene, cassiterite, columbite and tantalite.This indicates that they crystallised directly from the pegmatite melt. The presence of plagioclase and K-feldspar is evident for an intermediate Or-Ab composition that was subjected to subsolvus crystallisation under elevated water pressures of >5 kbar (Shelley 1993). The presence of sufficient amounts of water is further supported by the crystallisation of subhedral to euhedral petalite within the albitic units of the pegmatites. Petalite also indicates that the melt already was highly fractionated, enriched in incompatible elements as Li and in volatiles. Similar as for petalite, magmatic pyroxenes are interpreted to result from hydrous melts (France et al. 2013). Magmatic pyroxenes are represented by euhedral to subhedral spodumene. Crystallisation of oxide minerals took also place in the early stage of crystallisation. This is expressed by predominant euhedral to subhedral grains of cassiterite, columbite and tantalite as inclusions in or in interstitial positions between feldspar and quartz. Indeed, this also requires that the melt was saturated in Sn, Nb and Ta. Some studies (Linnen et al. 1992; Gomes and Neiva 2002) suggest that a part of the cassiterite can crystallise directly from the melt if a low salinity aqueous fluid is exsolved (Taylor and Wall 1992). Similar observations are made for the primary Nb and Ta oxide minerals columbite and tantalite that are cogenetic to cassiterite. According to van Lichtervelde et al. (2007) and Timofeev and Williams-Jones (2015), the heterogeneous internal compositions of these grains signal an overprint by multiple stages of hydrothermal alteration. However most authors suggested that vast portions of columbite and tantalite are of magmatic origin (Novak and Cerny 1998).

The following main stage is characterised by the crystallisation of the second generation of plagioclase, the growth of most of the muscovite and of quartz. The formation of the main mineral assemblage is accompanied by resorption and destruction of the early formed minerals, especially plagioclase, K-feldspar and spodumene,

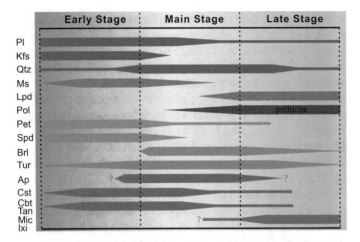

Fig. 3.6 Generalised crystallisation sequence for the LCT pegmatites at Bikita, Londonderry, Mount Deans, Cattlin Creek and Wodgina. Ap apatite; Brl beryl; Cbt columbite; Cst cassiterite; Ixi ixiolite; Kfs K-feldspar; Lpd lepidolite; Mic microlite; Ms muscovite; Pet petalite; Pl plagioclase; Pol pollucite; Qtz quartz; Spd spodumene; Tan tantalite; Tur tourmaline

as indicated by replacement structures. Textures like graphic intergrowth of quartz, feldspar and in part mica, developed during the main stage of crystallisation. According to experimental work, graphical intergrowth reflects disquilibrium growth (supersaturation) in melts that are vapour undersaturated and water bearing (MacLellan and Trembath 1991). Apatite and beryl occur as subhedral grains in interstitial position between the early crystallised minerals. Crystallisation of tourmaline continued during the main stage of crystallisation as indicated by more evolved tourmaline species (elbaitic, dravitic composition) and its textural position replacing spodumene. Some of the muscovite and beryl grains develop small rims enriched in Cs and Rb. The ongoing fractionation and enrichment of the melt with incompatible elements like Rb, Cs, Li and Be triggers the crystallisation of lepidolite that predominantly occurs within the inner portions of the pegmatite. The position of the massive pollucite mineralisation within the sequence of crystallisation stages at Bikita is difficult to ascertain (Fig. 3.2). On one hand, pollucite contains inclusions of petalite and feldspar, which signal that it crystallised during the early stage. On the other hand, pollucite is replaced by a prominent network of veins filled by lepidolite, quartz and petalite, indicating that it has formed prior to or contemporaneous to the onset of the pervasive lepidolite crystallisation. Therefore, the formation of the massive pollucite mineralisation is assigned to a late phase of the main stage crystallisation.

The late stage crystallisation (Fig. 3.6) is characterised by replacement reactions. This indicates that no more melt was involved. Hence, the mineralogical changes within this stage are attributed to circulation of deuteric hydrothermal fluids in a consolidated but still not fully cooled pegmatite. Replacement of plagioclase started along the cleavage plains or along small fissures and produced a fine grained intergrowth of mica and quartz. Lepidolite occurs in larger aggregates or masses in interstitial positions. K-feldspar was replaced by an intergrowth of bluish cleavelandite and lepidolite. Beryl is interpreted to be altered to secondary Be-minerals, like bavenite and/or bertrandite. Replacement of beryl started along small fractures. Along those fractures small rims and patchy zones are observed that are extremely enriched in Cs, representing pezzottaite. Tantalite and columbite are replaced by microlite, ixiolite and pyrochlor in irregularly shaped grains or complex aggregates. In massive mineralisation, pollucite was replaced by lepidolite, plagioclase, quartz and petalite along a crosscutting network of small veins. From BSE and EDX analysis it is evident that Cs was redistributed within the massive pollucite mineralisation and especially was concentrated along the contact to the vein network.

The early stage is characterised by subsolvus crystallisation from a hydrous melt that forms the predominant portions of plagioclase and K-feldspar contemporaneous to the crystallisation of petalite, spodumene, cassiterite and Nb–Ta oxide minerals. This early stage crystallisation shifts into the main stage of crystallisation accompanied by ongoing fractionation of the melt and the exsolution of a fluid phase. The main stage crystallisation is characterised by the growth of most of the muscovite and quartz as well as the occurrence of tourmaline, apatite and beryl. While muscovite is the predominant mica mineral at the beginning, it gets less abundant in the later main stage crystallisation where lepidolite dominates. From its structural position in the Bikita Main Quarry pegmatite and the textural observations it is suggested that the massive pollucite mineralisation formed during the main stage crystallisation. The late stage crystallisation of the LCT pegmatite systems is ruled by mineral replacement reactions. The mineralogical changes within this stage are attributed to deuteric hydrothermal fluids that circulated through a consolidated but still not fully cooled pegmatite system.

References

Dittrich T (2016) Meso- to Neoarchean Lithium-Cesium-Tantalum- (LCT-) pegmatites (Western Australia, Zimbabwe) and a genetic model for the formation of massive pollucite mineralisations. Dissertation Faculty of Geosciences, Geoengineering and Mining, TU Freiberg/Saxony, Germany, 341 pp. http://nbn-resolving.de/urn:nbn:de:bsz:105-qucosa-228968

France L, Ildefonse B, Koepke J (2013) Hydrous magmatism triggered by assimilation of hydrothermally altered rocks in fossil oceanic crust (northern Oman ophiolite). Geochem Geophys 14:2598–2614

Gomes MEP, Neiva AMR (2002) Petrogenesis of tin-bearing granites from Ervedosa, northern Portugal: the importance of magmatic processes. Chem Erde 72:47–72

Linnen RL, Williams-Jones AE, Martin RF (1992) Evidence of magmatic cassiterite mineralization at the Nong Sua aplite-pegmatite complex, Thailand. Can Mineral 30:739–761

MacLellan HE, Trembath LT (1991) The role of quartz crystallization in the development and preservation of igneous texture in granitic rocks: experimental evidence at 1 kbar. Am Mineral 76:1291–1305

Novák M, Cempírek J, Gadas P, Škoda R, Vašinová Galiová M, Pezzotta F, Groat LA (2015) Borosilite and Li, Be-bearing "Boron Mullite" $Al_8B_2Si_2O_{19}$, breakdown products of spodumene from the Manjaka Pegmatite, Sahatany Valley, Madagascar. Can Mineral 53:357–374

Novak M, Cerny P (1998) Niobium-tantalum oxide minerals from complex granitic pegmatites in the Moldanubicum, Czech Republic: primary versus secondary compositional trends. Can Mineral 36:659–672

Shelley D (1993) Igneous and metamorphic rocks under the microscope—classification, textures, microstructures and mineral preferred orientations. Chapman & Hall, London, p 445

Taylor JR, Wall VJ (1992) The behaviour of tin in granitoid magmas. Econ Geol 87:403–420

Thomas RJ, Bühmann D, Bullen WD, Scogins AJ, de Bruin D (1994) Unusual spodumene pegmatites from the Late Kibaran of southern Natal, South Africa. Ore Geol Rev 9:161–182

Timofeev A, Williams-Jones AE (2015) The origin of niobium and tantalum mineralization in the nechalacho REE deposit, NWT, Canada. Econ Geol 110:1719–1735

van Lichtervelde M, Salvi S, Bezait D (2007) Textural features and chemical evolution in tantalum oxides: magmatic versus hydrothermal origins for Ta mineralization in the Tanco Lower pegmatite, Manitoba, Canada. Econ Geol 102:257–276

Chapter 4
Geochemistry of LCT Pegmatites

The whole rock main and trace element geochemical data of the pegmatites is used to decipher the processes which led to the enormous Cs enrichment in the massive pollucite mineralisation. The geochemical study was also designed for potential prediction if a pegmatite potentially might host a yet not recognised massive pollucite mineralisation. Following these goals, the data from the pollucite-bearing Bikita pegmatite field are used for characterisation of the geochemical behaviour of Cs, and compared to the Western Australian pegmatites. The whole rock geochemical analyses are supplemented by the mineral modes and assemblages gained by automated SEM-MLA (Mineral Liberation Analysis) from the same samples. Representative slices of 211 samples were analysed for major (>1 wt%), minor (<1 wt% to >100 ppm) and trace (<100 ppm) elemental compositions. After crushing, milling and homogenising, an aliquot was used for embedding grains <500 µm in polished epoxy blocks for SEM-MLA study and another aliquot for main and trace element analyses by Australian Laboratory Services (ALS) Global laboratory in Rosia Montana (Romania), as detailed in the electronic appendix (Table A28). All data are afflicted by an inevitable intrinsic uncertainty due to the large grain sizes in pegmatites.

4.1 Major and Selected Trace Elements

There is a pronounced gap in Cs content between the pollucite bearing samples and all other mineral assemblages. This accentuates the solitary characteristic of pollucite. The pollucite samples contain 10–25 wt% Cs, whereas in the other mineral assemblages, upper limits of about 2–3 wt% Cs are not exceeded (Fig. 4.1). In the behaviour of Cs with respect to SiO_2 it is obvious that the pollucite bearing and mica dominated mineral assemblages tend to lower SiO_2 contents (40–50 wt%), whereas the other mineral assemblages contain about 60–80 wt% SiO_2. Variations

© The Author(s), under exclusive license to Springer Nature Switzerland AG 2019
T. Dittrich et al., *Archean Rare-Metal Pegmatites in Zimbabwe and Western Australia*,
SpringerBriefs in World Mineral Deposits, https://doi.org/10.1007/978-3-030-10943-1_4

within the SiO_2 content of the pollucite samples is explained by variable amounts of quartz and feldspar. In Cs versus Al_2O_3 coordinates, the pollucite, the mica dominated and the other mineral assemblages are separated groups. The Al_2O_3 content of pollucite (15–20 wt% Al_2O_3) is comparable to the other mineral assemblages that range between 10 and 25 wt% Al_2O_3. The mica dominated mineral assemblages are considerably more enriched in Al_2O_3 (25–30 wt%). The CaO contents are uniformly low and rarely exceed 1 wt% (Fig. A16). Pollucite exhibits low Na_2O of 2–3 wt%. Low Na_2O are also observed in samples with petalite and spodumene. The feldspar- and quartz dominated mineral assemblages exhibit a broad range from 4 to 11 wt% Na_2O (Fig. 4.1b). The K_2O exhibits an opposite behaviour compared to Na_2O (Fig. 4.1c). The K_2O concentrations are generally low within the quartz and feldspar dominated mineral assemblages (0–4 wt% K_2O) and high within the mica dominated mineral assemblages (6–12 wt% K_2O). Pollucite contains comparable values to the quartz and feldspar dominated mineral assemblages, whereas K_2O is almost absent within the spodumene and petalite bearing mineral assemblages. The Fe_2O_3 and MnO are uniformly low. Some samples with up to 20 wt% Fe_2O_3 (Fig. A16), and 1–2 wt% MnO from the Wodgina district can be treated as outliers and are apparently influenced by host rocks of BIF. To sum up, apart from the massive pollucite mineralisation at Bikita, similar main element trends with respect to the various mineral assemblages are observed from the pegmatites.

With the exception of the massive pollucite mineralisation at Bikita, the Li exhibits a positive correlation with Cs in most pegmatites. Lithium has its highest concentrations (1–3 wt%) within the mica dominated petalite and spodumene bearing mineral assemblages (Fig. 4.2). Pollucite and most of the quartz or feldspar dominated mineral assemblages have somewhat lower Li contents of about 0.1–1 wt%. Some of the samples display two distinct trends. One trend follows the positive correlation with increasing Cs, but there is also a negative correlation. As for Li, the Rb displays a strong positive correlation with increasing Cs, with highest concentrations (1–5 wt%) found in mica dominated mineral assemblages. The massive pollucite mineralisation at Bikita does not follow this trend, and Rb contents of 0.3–2 wt% as in most of the quartz and feldspar dominated mineral assemblages are observed there. The spodumene and petalite dominated mineral assemblages exhibit low Rb concentrations (30–200 ppm). In the Cs versus Sr coordinates (Fig. A17), the majority of the mica dominated and the petalite and spodumene dominated samples contain only traces of 1–10 ppm Sr. In the pollucite and the feldspar dominated mineral assemblages there are 10–100 ppm Sr. For Be the contents are low and range between 0.1 and 100 ppm. Mica dominant mineral assemblages exhibit slightly higher values than feldspar dominated mineral assemblages. The pollucite samples contain only a few ppm Be (1–5 ppm) within the lower range of all populations. Beryllium is almost absent within the petalite bearing mineral assemblages. No distinct differences exists between the various pegmatites.

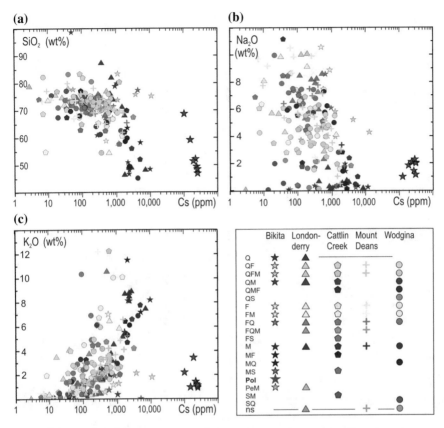

Fig. 4.1 Cs versus selected major oxides for defined mineral associations in LCT pegmatites. For abbreviation of mineral associations see Fig. 3.1. Prominent gap in Cs content between the pollucite bearing and the other mineral associations. **a** Cs versus SiO_2, with low SiO_2 for the mineral associations with pollucite and mica. Two pollucite samples with higher SiO_2 contents are due to higher amounts of quartz and petalite. **b** Cs versus Na_2O. Large variations in Na_2O content (<0.5–12 wt%) reflect primary mineralogy. Pollucite bearing mineral associations with low Na_2O (~2 wt%) are similar to most of the mica dominated mineral associations at <1 wt% Na_2O. **c** Cs versus K_2O. Large variations in K_2O content (<0.5–12 wt%) reflect the primary mineralogy. The pollucite bearing mineral association exhibits low K_2O (~1 wt%), comparable to most of the quartz and feldspar associations

The trends with Nb and Ta differ from those with the LILE. Quartz and most of the mica dominated mineral assemblages exhibit lower Nb and higher Ta contents between 1 and 1000 ppm (Fig. 4.2c). The pollucite samples display markedly lower Nb and Ta contents (1–20 ppm). Within the Li-bearing mineral assemblages MS and PeM, the Nb and Ta concentrations are even lower (0.3 ppm) or below detection limit, respectively. No marked differences are visible between the investigated LCT pegmatites (Fig. A17). The behaviour of Sn (Fig. 4.2d) at Bikita is similar to Nb and Ta. The lowest values are recorded for the petalite and spodumene bearing mineral

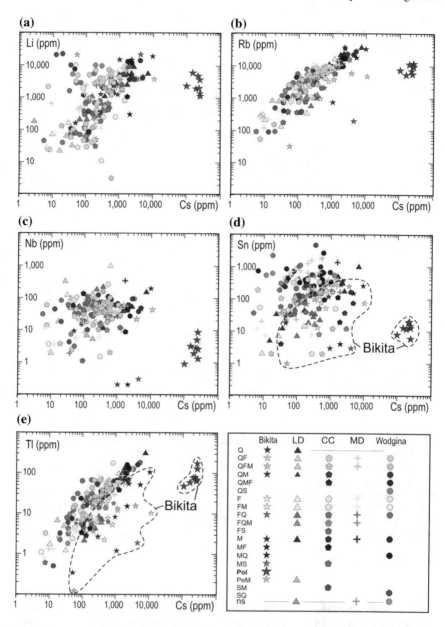

Fig. 4.2 Cs versus selected minor and trace elements in LCT pegmatites. For abbreviations of mineral associations see Fig. 3.1. Prominent gap in Cs content between the pollucite bearing mineral associations and the other samples. Samples from Bikita are summarised by broken lines. **a** Cs versus Li, with higher Li in pollucite-bearing associations. **b** Cs versus Rb, with high Rb contents in the pollucite and lepidolite bearing associations. **c–e** Cs versus Nb, Cs versus Tl, and Cs versus Sn, with low Nb, Sn and high Tl in pollucite bearing mineral associations

assemblages (1–5 ppm), followed by pollucite (5–30 ppm). The highest values are recorded for most of the feldspar, quartz and mica dominated mineral assemblages which contain between 30 and 500 ppm Sn. At the Wodgina pegmatite the Sn contents are higher (100–1000 ppm) and can even reach several thousands ppm. In samples from the Londonderry pegmatite field a positive correlation of Sn to Cs can be seen.

The element Tl is most enriched within the pollucite (40–200 ppm) and mica dominated (40–400 ppm) samples (Fig. 4.2e). Also Tl exhibits a pronounced positive correlation to Cs. The fluorine exhibits a positive correlation with Cs (Fig. A17). Among the pollucite samples a positive correlation of F to Cs can be seen, but displays a different trend towards higher Cs contents when compared to the other mineral assemblages. For chlorine a broad range between 50 ppm to about 700 ppm is observed from all mineral assemblages, except for the pollucite samples which have very low Cl of about 50 ppm.

In LCT pegmatite systems, fractional crystallisation is interpreted to cause economic enrichment of Cs, Li, Rb, Be, Nb, Ta and Sn. Certain elements are enriched within the crystal cumulus, when the mineral/melt distribution coefficients D are >1 (compatible elements) and thus their concentration is depleted within the remaining melt. When the values for D are <1 for distinct minerals, these incompatible elements are enriched in the melt. Numerous fractionation indicators are established to describe the development of melt during crystallisation. Examples are Al/Ga, Ba/Rb, Ba/Sr, Fe/Mn, K/Cs, K/Rb, K/Na, K/Tl, Mg/Li, Nb/Ta, Rb/Cs, Rb/Sr, Th/U or Zr/Hf (Kuester 2009; McKeough et al. 2013; Lima et al. 2014). A selection of such fractionation indicators is tested here to describe the melt development with respect to Cs in the pegmatite systems.

The Al/Ga decreases with increasing fractionation, as the incompatible Ga is enriched in the remaining melt/fluid, whereas Al is commonly incorporated into the structure of feldspars. At Bikita, no systematic relation between Al/Ga and Cs can be seen (Fig. 4.3a). The petalite and spodumene bearing mineral assemblages have slightly higher Al/Ga, whereas mica dominated mineral assemblages exhibit slightly lower ratios. This is in agreement with a decreasing Al/Ga ratio during fractional crystallisation. The pollucite samples exhibit comparable Al/Ga ratios of 2500–5000 as most of the feldspar dominated mineral assemblages. This indicates that pollucite crystallised either in an early step of fractional crystallisation or was formed by another, not yet discovered process. At the Londonderry and Mount Deans pegmatite fields, the samples follow a general trend from higher Al/Ga ratios (>1000) within the quartz and feldspar dominated mineral assemblages towards lower Al/Ga ratios (<1000) in the mica dominated mineral assemblages. This trend is further accompanied by an increasing Cs content and is interpreted to reflect common fractional crystallisation. In contrast, the samples from the Wodgina and Cattlin Creek pegmatite stagnate around Al/Ga ratios of 2000, indicating that fractional crystallisation was not the dominating process during consolidation of these pegmatites.

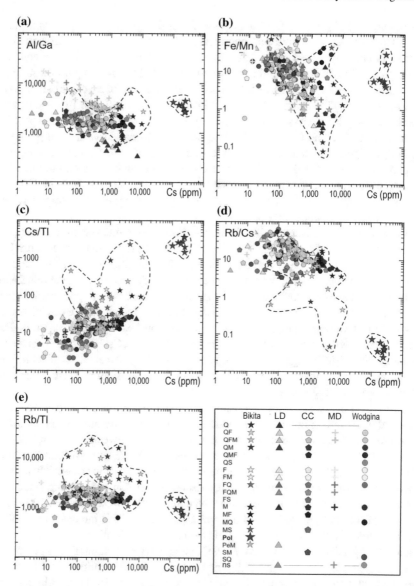

Fig. 4.3 Cs versus selected fractionation indicators in LCT pegmatites. For abbreviation of mineral associations see Fig. 3.1. Samples from Bikita are summarised by broken lines. Prominent gap in Cs content between the pollucite mineral association and remaining samples. **a** Cs versus Al/Ga: Broad range in Al/Ga ratios; Bikita, Londonderry and Mount Deans indicate a weak trend towards lower Al/Ga ratios. **b** Cs versus Fe/Mn: Broad variation in Fe/Mn ratios indicating a general fraction trend towards lower Fe/Mn ratios in the highest evolved portions. **c** Cs versus Cs/Tl: Broad range, with highest values at Bikita. **d** Cs versus Rb/Cs: Pollucite bearing samples display the lowest Rb/Cs ratios. **e** Cs versus Rb/Tl: Most samples from Bikita display the highest Rb/Tl. Pollucite bearing samples show similar ratios as the other samples

The Mn concentration will increase during ongoing fractional crystallisation, causing a lowering Fe/Mn ratio (Bogoch et al. 1997). At the Bikita pegmatite field (Fig. 4.3b) this is indicated by higher Fe/Mn ratios (>1) in feldspar and quartz dominated mineral assemblages, whereas mica dominated mineral assemblages exhibit the lowest Fe/Mn ratios (1–0.1). Pollucite samples have comparable Fe/Mn ratios (4–8) to most of the feldspar bearing mineral assemblages. This indicates that pollucite crystallises relatively early within the course of fractional crystallisation or was precipitated due to a different process. The two pollucite mineral assamblage samples that exhibit higher Fe/Mn ratios contain variable amounts of other minerals. The Western Australian LCT pegmatites exhibit a similar negative correlation of the Fe/Mn ratio and Cs. Nevertheless, only some samples of the feldspar and mica dominated mineral assemblages from the Londonderry pegmatite field and Wodgina pegmatite reach degrees of fractionation comparable to the highest evolved mica mineral assemblages of the Bikita pegmatite field.

Cs and Tl appear as incompatible elements. They display a well pronounced positive correlation. The Tl is more incompatible with respect to the typical pegmatite mineral assemblages than Cs. The samples from Bikita exhibit a broad range but in general have the highest Cs/Tl (30–30,000) among all pegmatites (Fig. 4.3c). This may serve as an indicator for hidden massive pollucite mineralisation. Samples from the Londonderry pegmatite field describe two distinct trends for the Cs/Tl ratio: The first trend is characterised by most of the mica dominated mineral assemblages and describes a negative correlation with a smooth slope, whereas the second trend describes a negative correlation with a steeper slope and is characterised by the feldspar dominated mineral assemblages. These two trends are also present at Bikita. This may indicate that at least two different processes were active during the crystallisation of the pegmatites. Samples from the Cattlin Creek and Wodgina, as well as from the Mount Deans pegmatite field describe negative correlation, but in general follow the first trend described for the Londonderry pegmatite field.

The Rb/Cs data obtained from Bikita exhibit the broadest range from about 80 within quartz and feldspar dominated mineral assemblages to 0.09–0.02 for the pollucite and spodumene bearing mineral assemblages (Fig. 4.3d). The mica dominated mineral assemblages exhibit comparable higher Rb/Cs ratios (3–30) as the feldspar dominated mineral assemblages. A large gap in ratios between the pollucite and spodumene bearing samples and most of the other samples is observed. The Cs versus Rb/Tl plot displays a very significant contrast between Bikita and the Western Australian pegmatites. Samples from Bikita display a pronounced negative correlation with the Rb/Tl that range from 3000 to about 50, whereas the pollucite samples and the spodumene bearing mineral assemblage exhibiting the low Rb/Tl of 50–200 (Fig. 4.3e). In all the plots, a marked gap exists between the pollucite and the other mineral assemblages.

4.2   Rare Earth Elements

The rare earth element (REE) distribution pattern in igneous rocks and minerals are mainly controlled by the ion charge (3+) and the decreasing ionic radii with increasing atomic number. Fractionation of LREE (light rare earth elements) and HREE (heavy rare earth elements) from case to case are induced by the small differences among the ionic radii and the variations of the diverse mineral/melt (D) values. The REE data are usually presented as values normalised to a reference chondrite (McDonough and Sun 1995). The REE contents in the pegmatites turned out to be generally very low, and even below the detection limit for a large number of analyses. Only samples with complete REE pattern, except for Eu with large negative anomaly are presented here.

The REE in Bikita pegmatite are low (Fig. 4.4a). The LREE are slightly more enriched than in comparison to the HREE. Some samples display a pronounced negative Eu anomaly and almost all samples are characterised by a small negative Ce anomaly. The pollucite sample (DS01) is depleted in most REE compared to the chondrite, but matches the pattern of the other samples. Two types of REE patterns are recognised for the Londonderry pegmatite field. Type-I exhibits higher LREE compared to HREE and exhibits a pronounced negative Eu anomaly (Fig. 4.4b). The type-II REE pattern is characterised by a discontinuous distribution of the REE (Fig. 4.4c). The LREE and HREE are generally low, with the HREE elements Gd to Dy being slightly more enriched than the LREE. Several samples are characterised by a positive Tm anomaly, possibly due to analytical interferences. The most prominent features are enrichment of Gd and Tb, as well as in parts of Dy and Sm. Samples from the Mount Deans pegmatite field display three distinct type of REE pattern. Type-I is characterised by enrichment of LREE compared to HREE. Europium and Ce anomalies are absent or only weakly negative. The type-II displays a discontinuous pattern with a general enrichment of LREE compared to HREE, but exhibits a pronounced negative Eu anomaly and distinct higher values of Sm, Gd and Tb. The type-III of REE pattern is unique to the Mount Deans pegmatite field and characterised by a general enrichment of HREE compared to LREE (Fig. 4.4d). The samples from the Cattlin Creek pegmatite exhibit two distinct types of REE patterns. Type-I exhibits a variable higher values of LREE compared to HREE (Fig. A18). Type-II patterns are discontinuous, with a general enrichment of LREE compared to the HREE and exhibit a pronounced negative Eu anomaly. Furthermore, they display higher contents of Sm, Gd, Tb and Tl. At Wodgina the type-I pattern exhibits slightly higher LREE compared to HREE (Fig. 4.4e). However, in contrast to Bikita, Londonderry and Cattlin Creek, the type-I pattern at Wodgina exhibits a pronounced positive Eu anomaly (Fig. A18). The type-II displays a discontinuous pattern with still a slight enrichment of LREE compared to HREE, but exhibits a pronounced negative Eu anomaly and distinct higher values for Sm, Gd, Tb and Dy (Fig. 4.4f). The presence of two distinct REE pattern types suggests that two petrological processes were involved during the formation of the LCT pegmatites.

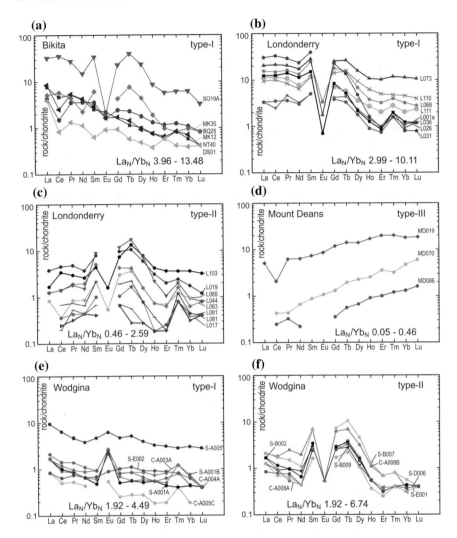

Fig. 4.4 Chondrite normalized REE distribution in LCT pegmatites; normalization values according to McDonough and Sun (1995). **a** Type-I REE pattern of selected samples from Bikita. LREE are enriched when compared to HREE. **b** Type-I REE pattern of selected samples from the Londonderry pegmatites. **c** Type-II discontinuous REE pattern in the Londonderry pegmatites. Eu displays negative anomaly or is below detection limit. Several samples show strong positive Tm anomaly. **d** Type-III REE distribution in some samples from Mount Deans pegmatites. **e** Type-I REE pattern of selected samples from the Wodgina pegmatites. **f** Type-II REE pattern of selected samples from the Wodgina pegmatites

References

Bogoch R, Bourne J, Shirav M, Harnois L (1997) Petrochemistry of a late Precambrian garnetiferous granite, pegmatite and aplite, southern Israel. Mineral Mag 61:111–122

Kuester D (2009) Granitoid-hosted Ta mineralization in the Arabian-Nubian Shield: Ore deposit types, tectono-metallogenetic setting and petrogenetic framework. Ore Geol Rev 35:68–86

Lima SM, Neiva AMR, Ramos JMF, Cuestad A (2014) Long-lived magmatic systems and implications on the recognition of granite-pegmatite genetic relations: characterization of the Pavia-granitic pegmatites (Ossa-Morena Zone, Portugal). Chem Erde 74:625–639

McDonough WF, Sun SS (1995) The composition of the earth. Chem Geol 120:223–253

McKeough MA, Lentz DR, McFarlane CRM, Brown J (2013) Geology and evolution of pegmatite-hosted U-Th REE-Y-Nb mineralization, Kulyk, Eagle, and Karin Lakes region, Wollaston Domain, northern Saskatchewan, Canada: examples of the dual role of extreme fractionation and hybridization processes. J Geosci 58:321–346

Chapter 5
Geochronology of Archean LCT Pegmatites

It is appropriate to apply several age dating methods to the pegmatites, as each method addresses different processes. Zircon and monazite can be juvenile in the pegmatite forming melt, but may also crystallise during later hydrothermal and metamorphic events. Furthermore zircon and monazite crystals can be xenocrysts from the host rocks or the source rock region during the ascent of pegmatite melt through the crust. Age dating of those crystals will yield older inherited ages that are not related to the crystallisation of the pegmatite. Dating of zircon had to be discarded in this study due to the lack of suitable grains in the entire sample set. The U–Pb dating of Ta–Nb–Sn oxides and cassiterite by LA-ICPMS allows the determination of the ages of the main crystallisation. This method turned out to be suitable for the Neo-Archean pegmatites due to the widespread occurrence of these minerals as accessories in almost all samples. They are formed at a virtually constant level during pegmatite crystallisation. Furthermore, the Nb- and Ta-oxides are present in at least two generations as indicated by the overgrowth of columbite and tantalite by microlite or ixiolite. Mineral separates of Nb- and Ta-oxides, and cassiterite were produced using standard separation and concentration techniques. Polished blocks with embedded mineral grains were prepared and 25 samples analysed by LA-ICPMS. The measurements were conducted at the facilities of the Department of Petrology and Geochemistry at the Goethe University Frankfurt/Main, using a Thermo-Scientific Element II sector field ICPMS, coupled to a New Wave UP213 ultraviolet laser system (Text A27).

The Th-U-total Pb method on monazite by the electron microprobe was applied. Monazite is an accessory mineral that occurs in felsic magmatic and Al-rich metamorphic rocks and is also known from many pegmatites worldwide. The dating method is based on the premise that all Pb in monazite is radiogenic, generated by the decay of Th and U. Monazite apparently incorporates only negligible amounts of common Pb during crystallisation, and allows only extremely low diffusion for Pb (Montel et al. 1996). Experimental work by Seydoux-Guillaume et al. (2002) has shown that fluid controlled recrystallisation can reset the monazite chronometer.

© The Author(s), under exclusive license to Springer Nature Switzerland AG 2019
T. Dittrich et al., *Archean Rare-Metal Pegmatites in Zimbabwe and Western Australia*,
SpringerBriefs in World Mineral Deposits, https://doi.org/10.1007/978-3-030-10943-1_5

In contrast to Panafrican pegmatites, the REE-phosphate monazite turned out to be rare within the Neo-Archean pegmatites and was only found in a single sample at Londonderry.

The ^{40}Ar–^{39}Ar method on mica generally yields cooling ages (Harrison et al. 2009) and thus will date the latest stage of the crystallisation or consolidation of the LCT pegmatites. The ^{40}Ar–^{39}Ar-dating was performed at the Argon Laboratory Freiberg (ALF) based at the Institute of Geology at the TU Bergakademie Freiberg/Saxony. Muscovite and lepidolite from 16 samples were handpicked and analysed by mass spectrometry. Raw data were treated with a laboratory intern MatLab toolbox. Finally, PA (plateau age), TFA (total fusion age), IA (isochron age) and IIA (inverse isochron) ages were calculated according to Ludwig (2003). For some samples it was possible to calculate the plateau ages using almost the complete gained ^{39}Ar fraction. For samples, with Ar loss or excess Ar, only the arbitrary undisturbed steps are used for the determination of the plateau ages.

5.1 Bikita Pegmatite Field

A total of 19 grains were analysed by 69 single spot analyses in sample BQ20 by U–Pb LA-ICPMS. Individual spots from this sample represent predominantly cassiterite, tantalite (Mn, Fe) and minor ixiolite, microlite or fine intergrowths of these phases. One analysis was performed on a single zircon-microlite intergrowth. The U/Pb concordia diagram yields an upper intercept age of 2616 ± 16 Ma (Fig. 5.1a). A subset of 17 grains from sample BQ25 were analysed in 47 individual spots, predominantly tantalite and Mn-columbite, with subordinate cassiterite and minor ixiolite. The upper intercept age is 2625 ± 11 Ma (Fig. 5.1b). Further six grains were analysed in 17 spots on sample MK33b that predominantly contains Fe-columbite with only minor Mn-columbite and pyrochlor. Data in the U/Pb concordia diagram illustrates an upper intercept age of 2629 ± 9 Ma. The upper intercept U/Pb ages from the Bikita pegmatite field range within their respective errors and are interpreted as crystallisation ages of the Ta–Nb–Sn oxide minerals. The data confirm the 2617 ± 1 Ma age obtained from columbite by Melcher et al. (2015). Furthermore, this age coincides with the intrusion of the Chilimanzi Suite of granites that was emplaced at 2601 ± 14 Ma (Jelsma et al. 1996).

Three samples from the Bikita Main Quarry pegmatite were dated by the Ar–Ar method (Fig. 5.2a, b). One sample (BIK01B) from the border zone of the pegmatite, and a lepidolite separated from the lepidolite bearing veins from the massive pollucite mineralisation (BIKb, BQ01-Lpd) yielded minimum ages ranging from 2376 to 2100 Ma. The Ar isotopic system of mica was reopened and reset during an event in the Paleoproterozoic. Even though, this event is not accompanied by an ascertainable tectonic reworking of the pegmatite. The established closing temperatures of mica (~350–425 °C; Harrison et al. 2009) would suggest a considerable reheating of the Bikita pegmatite ~200 Ma after its crystallisation. Furthermore, this event is interpreted to be related to pervasive late stage replacement reactions. The Paleopro-

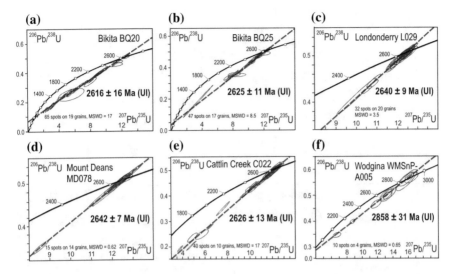

Fig. 5.1 Results of U–Pb dating by LA-ICPMS of Ta-, Nb- and Sn-oxide minerals in LCT pegmatites (selected samples). **a, b** Bikita pegmatite. **c** Londonderry pegmatite. **d** Mount Deans pegmatite. **e** Cattlin Creek pegmatite. **f** Wodgina pegmatite

terozoic age of this thermal overprint is distinctively younger than the intrusion of the Great Dike and associated rocks at 2575 Ma (Soederlund et al. 2010; Oberthuer et al. 2002).

5.2 Londonderry Pegmatite Field

The analysis of sample L001 with 25 spot measurements from six grains with Fe- and Mn-columbite, and only minor areas displaying pyrochlor compositions yields an upper intercept age of 2676 ± 24 Ma. Sample L029 (20 grains of Fe-columbite, 31 spots) displays an age of 2640 ± 9 Ma (Fig. 5.1c). Further 12 grains of Mn-columbite in sample L048 were analysed in 25 spots and yield an age of 2627 ± 7 Ma. Sample L063 with 12 grains of ixiolite (34 spots), one cassiterite (8 spots) and one Mn-tantalite (6 spots) gives an upper intercept age at 2624 ± 7 Ma. Sample L108 (21 grains of ixiolite) yields an upper intercept age of 2633 ± 8 Ma. Thus, with the exception of sample L001 with 2676 ± 24 Ma, the Nb-Ta minerals at the Londonderry pegmatites crystallised at about 2630 Ma (Dittrich 2016; Dittrich et al. 2017). The age range from 2640 to 2620 Ma from the pegmatites correlates with a narrow event during M3b metamorphism, which is characterised by a second period of high heat flux into the upper crust, with the emplacement of a low-Ca granitoid suite, and also to the 2630 Ma event of Au mineralisation (Fig. 2.7).

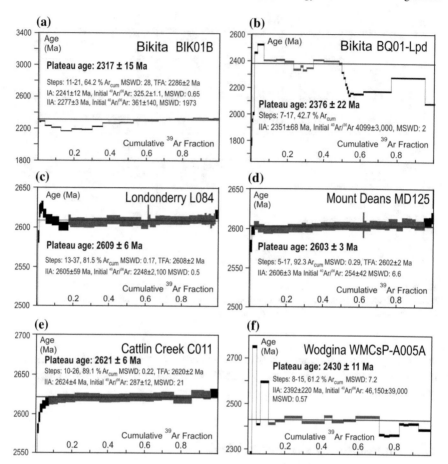

Fig. 5.2 Results of ^{40}Ar/^{39}Ar dating of mica minerals in LCT pegmatites (selected samples). Steps used for the calculation of plateau age are red. Rejected steps are in black. Blue line indicates the weighted plateau age. Box heights correspond to 1σ. **a, b** Lepidolite from Bikita. **c** Muscovite from Mount Deans. **d** Lepidolite from Londonderry. **e** Muscovite from Cattlin Creek. **f** Lepidolite from Wodgina

Five representative mica samples from the Londonderry pegmatite field were dated by the Ar-Ar method with small heating steps. Most measured samples exhibit good plateaus and thus are interpreted as reliable cooling ages of the micas. With the exception of the age of 2625 ± 7 Ma, obtained from sample L036, all other ages scatter within the range of error around 2605 Ma, with sample L084 at 2609 ± 6 Ma (Fig. 5.2c).

Monazite in Londonderry pegmatites occurs as inclusion within beryl. The crystals range up to 3 mm in size, with no inclusions, and in part exhibit a concentrically or patchy zonation. Monazite is also present in hypidiomorphic to xenomorphic crystals in interstitial position to tantalite and columbite along fractures in beryl, or along

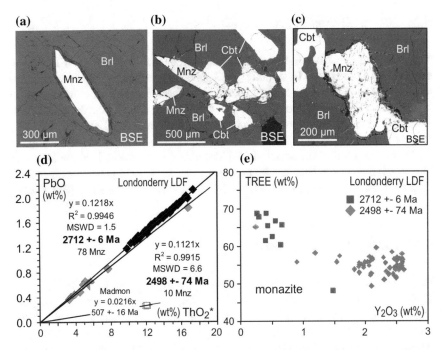

Fig. 5.3 Electron microprobe Th-U-Pb monazite dating in Londonderry pegmatite. **a** Isolated monazite (Mnz) in beryl (Brl). **b, c** Monazite (Mnz) intergrown with columbite (Cbt) and beryl. **d** Total PbO versus ThO₂* isochrone diagram. Two ages of populations at 2712 ± 6 and at 2498 ± 74 Ma are weighted average ages (Ludwig 2003) calculated from analyses defining isochrone regressions forced through zero (Montel et al. 1996). **e** Variation of Y₂O₃ and TREE with ages in monazite

grain boundaries between beryl, feldspar, mica or quartz crystals. These monazites are further characterised by many inclusions (Fig. 5.3a–c). Quantitative analyses of monazite grains by electron microprobe were used to calculate Th–U–Pb ages (Fig. 5.3d). The UO₂ is assigned to ThO₂ equivalents labelled as ThO₂*. Isochron ages from regression in the ThO₂* versus PbO diagram coincide well with weighted average ages calculated using Isoplot 3.0 (Ludwig 2003). Two distinct generations of monazites are deciphered. A well pronounced older age dated at 2712 ± 6 Ma (2727 ± 14 Ma in sample L068), and a less pronounced younger generation dated at 2498 ± 74 Ma. The existence of the two monazite generations is further supported by different mineral chemistry (Fig. 5.3e). The older age may be interpreted as inherited monazite. Although only poorly defined, the younger age suggests an overprinting process.

5.3 Mount Deans Pegmatite Field

Three samples with Ta–Nb–Sn-oxide minerals were analysed. Sample MD078 was analysed by 12 spots of Mn-tantalite and 3 spots on cassiterite. The upper intercept age is 2642 ± 7 Ma (Fig. 5.1d). Data from further 19 grains of cassiterite from sample MD125 yield an age of 2638 ± 8 Ma. Sample MD 131 with predominant cassiterite (12 spots) and minor Mn-tantalite (2 spots) was dated at 2639 ± 22 Ma. In summary, a relative narrow time interval around 2640 Ma for the crystallisation of the Ta–Nb–Sn-oxide minerals within the Mount Deans pegmatite field can be stated. One lepidolite sample from the Mount Deans pegmatite field yields a well-defined plateau at 2603 ± 3 Ma and is interpreted to represent the cooling age (Fig. 5.2d).

5.4 Cattlin Creek Pegmatite

Five samples from the Cattlin Creek pegmatite were analysed by U–Pb dating of Ta–Nb–Sn-oxide minerals. Sample C001 with 8 grains of cassiterite (13 spots) and Mn-columbite (7 spots), as well as one analysis from ixiolite and a complex intergrowth of columbite and tantalite yields an age of 2630 ± 20 Ma. A total of 28 spots within 20 grains of Mn-columbite in sample C011 allowed to calculate an upper intercept age at 2625 ± 14 Ma. Sample C016a (5 grains) of microlite (4 grains, 16 spots) and subordinate zircon (1 grain, 5 spots) yields an age of 2567 ± 49 Ma. The analysis of sample C022 comprises 10 grains of Mn-columbite (23 spots) and Mn-tantalite (9 spots), microlite (3 spots) and tantalite (2 spots). The resulting upper intercept U-Pb age is 2626 ± 13 Ma (Fig. 5.1e). Further 20 grains were analysed in 45 spots from sample C055 which consists of Mn-columbite (22 spots), Mn-tantalite (13 spots), microlite (6 spots), ixiolite (2 spots), zircon (1 spot) and cassiterite (1 spot). The resulting age is calculated at 2616 ± 16 Ma. With the exception of sample C016 with its relatively young age of 2567 ± 49 Ma and a large error, the other samples define a relative narrow time interval of crystallisation from 2630 to 2615 Ma (Dittrich 2016; Dittrich et al. 2017).

Four mica samples from the Cattlin Creek pegmatites contain white mica (samples C010, C011) and lepidolite (samples C016, C040). Plateau age data display a different behaviour of these minerals during degassing. The white mica samples C010 and C011 display well developed plateaus and yield identical ages of 2624 ± 7 and 2621 ± 6 Ma (Fig. 5.2e). The degassing behaviour of the lepidolite samples is different. Samples C016 and C040 exhibit disturbed age spectra and actually reveal younger, but identical ages of 2599 ± 6 and 2598 ± 6 Ma. Consequently, the complete cooling and closure of the Ar isotope system within the Cattlin Creek pegmatite lasted for at least about 20–25 Ma, starting at ~2625–2600 Ma. This time range is interpreted to have been caused by the different closure temperatures for muscovite and lepidolite. The crystallisation of muscovite took place relatively early during the formation, whereas lepidolite was formed in the last stages of the development of the Cattlin Creek pegmatite.

5.5 Wodgina Pegmatite

From Wodgina, four samples from the Mount Tinstone open pit and two samples from the Mount Cassiterite open pit were dated by the U–Pb LA-ICPMS method on Ta–Nb–Sn oxides. The Mount Tinstone sample WMSnP-A005 (4 grains) with cassiterite (9 spots) and microlite (1 spot) yields an upper intercept age of 2858 ± 31 Ma (Fig. 5.1f). Another 19 grains from sample WMSnP-B007 with Mn-tantalite (39 spots), cassiterite (9 spots), microlite (4 spots) and ixiolite (3 spots) was calculated at 2863 ± 12 Ma. Sample WMSnP-B011a consists of 19 grains of Mn-tantalite (29 spots) with only minor ixiolite (2 spots) which give an upper intercept age of 2865 ± 22 Ma. The 25 spot analysis from 19 grains of cassiterite and ixiolite in sample WMSnP-B011b yield an age of 2849 ± 24 Ma. At the Mount Cassiterite open pit, sample WMCsP-A006a, with five grains of Mn-tantalite (6 spots) and ixiolite (3 spots) yields an age of 2871 ± 23 Ma which is similar to data from the Mount Tinstone open pit. In sample WMCsP-A004b an age of 2711 ± 310 Ma was calculated from six grains of microlite (6 spots), ixiolite (6 spots) and cassiterite with 2 spots (Fig. A19; Table A20). A significant shift for most of the analysed spots to the lower intercept of the U–Pb concordia signals a Pb-loss during a later event. Apart from this deviating sample, at the Wodgina pegmatite a narrow age range from 2870 to 2850 Ma is observed for the formation of the Ta–Nb–Sn oxide minerals (Dittrich 2016; Dittrich et al. 2017). This age span is roughly within the range of the world first minor LCT pegmatite formation event at 2850 to 2800 Ma. The age span age coincides further with the emplacement of two major magmatic supersuites, the Cutinduna Supersuite at 2910–2895 Ma, and the Split Rock Supersuite at 2890–2830 Ma (van Kranendonk et al. 2006).

Five lepidolite samples from the Mount Cassiterite and Mount Tinstone pegmatites were investigated. by the Ar–Ar method. As already observed from Bikita, none of the investigated samples from Wodgina exhibits a well defined plateau age. The obtained ages range from ~2600 to 2300 Ma, and thus are comparable to the ages of the earlier work of Jeffrey (1956) who dated muscovite and microcline by means of the K–Ar-method at 2420 and 2220 Ma (Fig. 5.2f). However, these ages are about 200–500 Ma younger than the ~2800 Ma U–Pb SHRIMP ages obtained by Kennedy (1998) and Kinny (2000) on apatite and tantalite. Therefore, the obtained mica cooling ages of 2600 to 2300 Ma suggest that the Ar isotope system was subjected to a later hydrothermal event, or was kept open for an extreme long period of time. The Ar–Ar ages postdate the final assembly of the Pilbara Craton at about 2800 Ma (van Kranendonk et al. 2007), but are significant older than collision of the Yilgarn and Pilbara cratons during the Opthalmia Orogeny and Capricorn Orogeny at 2215–2145 Ma (Rasmussen et al. 2005) and 1830–1780 Ma (Cawood and Tyler 2004), respectively.

References

Cawood PA, Tyler IM (2004) Assembling and reactivating the Proterozoic Capricorn Orogen; lithotectonic elements, orogenies, and significance. Precambr Res 128:201–218

Dittrich T (2016) Meso- to Neoarchean Lithium-Cesium-Tantalum- (LCT-) pegmatites (Western Australia, Zimbabwe) and a genetic model for the formation of massive pollucite mineralisations. Dissertation Faculty of Geosciences, Geoengineering and Mining, TU Freiberg/Saxony, Germany, 341 pp. http://nbn-resolving.de/urn:nbn:de:bsz:105-qucosa-228968

Dittrich T, Seifert T, Schulz B, Pfänder J, Gerdes A (2017) Formation of LCT pegmatites in archean cratons: constraints from $^{40}Ar/^{39}Ar$ Mica, U–Th–Pb Monazite and U–Pb Tantalite/Columbite dating. Goldschmidt 2017 Conference, Paris, Abstracts, p 959

Harrison TM, Célérier J, Aikman AB, Hermann J, Heizler MT (2009) Diffusion of ^{40}Ar in muscovite. Geochim Cosmochim Acta 73:1039–1051

Jeffrey PM (1956) The radioactive age of four Western Australian pegmatites by the potassium and rubidium methods. Geochim Cosmochim Acta 10:191–195

Jelsma HA, Vinyu ML, Valbrachet PJ, Davies GR, Wijbrans JR, Verdurmen EAT (1996) Constraints on Archaean crustal evolution of the Zimbabwe craton: A U–Pb zircon, Sm–Nd and Pb–Pb whole-rock isotope study. Contrib Mineral Petrol 124:55–70

Kennedy AK (1998) SHRIMP ages of apatites from Pilbara tin-bearing pegmatites. Geol Soc Austral Abstracts 49:242

Kinny PD (2000) U–Pb dating of rare metal (Sn–Ta–Li) mineralized pegmatites in Western Australia by SIMS analysis of tin and tantalum bearing ore minerals. New Frontiers in Isotope Geoscience Abstracts and Proceedings Lorne (Australia), pp 113–116

Ludwig KR (2003) User manual for isoplot/ex. rev. 2.49.2. a geochronological toolkit for microsoft Excel. Berkeley Geochronological Center Spec Publ 1a, 55 pp

Melcher F, Graupner T, Gäbler HE, Sitnikova M, Henjes-Kunst F, Oberthuer T, Gerdes A, Dewaele S (2015) Tantalum-(niobium-tin) mineralisation in African pegmatites and rare metal granites: constraints from Ta–Nb oxide mineralogy, geochemistry and U–Pb geochronology. Ore Geol Rev 64:667–719. https://doi.org/10.1016/j.oregeorev.2013.09.003

Montel JM, Foret S, Veschambre M, Nicollet C, Provost A (1996) Electron microprobe dating of monazite. Chem Geol 131:37–53

Oberthuer T, Davis DW, Blenkinson TG, Höhndorf A (2002) Precise U–Pb mineral ages, Rb–Sr and Sm–Nd systematics for the Great Dyke, Zimbabwe—constraints on late Archean events in the Zimbabwe craton. Precambr Res 113:293–305

Rasmussen B, Fletcher IR, Sheppard S (2005) Isotopic dating of the migration of a low-grade metamorphic front during orogenesis. Geology 33:773–776

Seydoux-Guillaume AM, Paquette JL, Wiedenbeck M, Montel JM, Heinrich WH (2002) Experimental resetting of the U–Th–Pb systems in monazite. Chem Geol 191:165–181

Soederlund U, Hofmann A, Klausen MB, Olsson JR, Ernst RE, Persson PO (2010) Towards a complete magmatic barcode for the Zimbabwe craton: baddeleyite U–Pb dating of regional dolerite dyke swarms and sill complexes. Precambr Res 183:388–398

van Kranendonk MJ, Hickman AH, Smithies RH, Williams IR, Bagas L, Farrell TR (2006) Revised lithostratigraphy of Archean supracrustal and intrusive rocks in the northern Pilbara Craton, Western Australia. Geol Surv West Austral Record 2006–15, 63 pp

van Kranendonk MJ, Smithies RH, Hickman AH, Champion DC (2007) Review: secular tectonic evolution of Archean continental crust: interplay between horizontal and vertical processes in the formation of the Pilbara Craton, Australia. Terra Nova 19:1–38

Chapter 6
Radiogenic and Stable Isotopes, Fluid Inclusions

6.1 Whole Rock Sm–Nd Isotope Compositions

Samarium and Nd isotope data are used to provide information on the source of melts, as well as to determine the age of the rocks. Main advantage of the Sm–Nd isotope method is the geochemical behaviour of the Sm and Nd which apparently are not affected by partitioning during melting. In contrast, fractionation during crystallisation of minerals then influences the Sm/Nd ratio of the material. The radioactive decay of the long-living ^{147}Sm isotope (half life: $1.06 * 10^{11}$ Ma) to the radiogenic ^{143}Nd isotope change the ^{143}Nd/^{144} Nd ratio of the considered rock. Fundamental Sm–Nd isotope investigations by DePaolo and Wasserburg (1976) and DePaolo (1981) could demonstrate that various reservoirs within the earth exhibit different values of initial ^{143}Nd/^{144}Nd ratios. Thus, if the age of the investigated igneous rock is known, the initial ^{143}Nd/^{144}Nd ratio can provide information when the former melt was separated from the mantle. This depleted mantle model age then is useful for the magmatogenetic interpretation of the samples.

Fifteen whole rock pegmatite samples (3 from each pegmatite) were investigated for the Sm and Nd isotopic composition. Sample preparation and analysis were performed at the laboratory for mass spectrometry at the Goethe University/Frankfurt am Main (Germany) and summarised in Dittrich (2016), and presented in Table A21. Several samples yield extreme high ^{147}Sm/^{144}Nd ratios of >0.3, indicating extreme fractionation of the REE. This is also reflected in their calculated present day εNd values of $εNd_{t=0}$ at >40.0. The highest ^{147}Sm/^{144}Nd and εNdt = 0 are recognised for sample BIK01B (^{147}Sm/^{144}Nd = 0.7966, $εNd_{t=0}$ = 200.1) and WMSnP-E001 (^{147}Sm/^{144}Nd = 0.931, $εNd_{t=0}$ = 272.8). The reason for the high degree of REE fractionation remains ambiguous but is suggested either to be related to the late stage replacement processes and related element redistribution (e.g., feldspar replacement, replacement of the primary Nb–Ta–Sn-oxides), or to reflect the mineralogical composition of the samples.

© The Author(s), under exclusive license to Springer Nature Switzerland AG 2019
T. Dittrich et al., *Archean Rare-Metal Pegmatites in Zimbabwe and Western Australia*, SpringerBriefs in World Mineral Deposits, https://doi.org/10.1007/978-3-030-10943-1_6

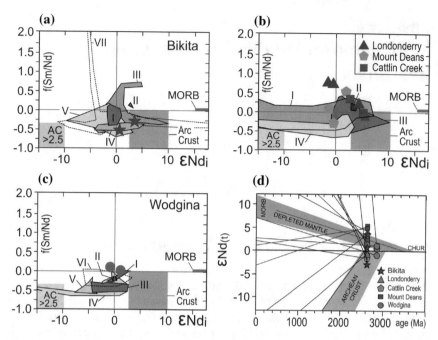

Fig. 6.1 Isotopic composition of initial εNd$_i$ versus fractionation factor f(Sm/Nd) of Roddaz et al. (2007) for identification of magmatic sources of LCT pegmatites. Data for sources for each craton are compiled from numerous athors. AC > 2.5—Archean Crust older than 2.5 Ga. **a** Bikita and potential sources from the Zimbabwe Craton: I—greenstone belts; II—komatiites; III—mafic dykes; IV—felsic igneous rocks; V—TTG (trondhjemite-tonalite-gneisses); VI—metasedimentary rocks. **b** Pegmatites from the Yilgarn Craton and potential sources from the Yilgarn Craton: I—greenstone belt lithologies; II—komatiites; III—basaltic rocks; IV—felsic magmatic rocks and gneiss. **c** Wodgina pegmatite and potential sources from the Pilbara Craton: I—greenstone belts; II—mafic rocks; III—basaltic rocks; IV—felsic magmatic rocks; V—ademellites; VI—metasedimentary rocks. **d** Isotopic composition (εNd) versus time for whole rock pegmatite samples and potential source regions. CHUR—DePaolo and Wasserburg (1976); DM depleted mantle—DePaolo (1981); MORB mid ocean ridge basalt, Archean crust—Milisenda et al. (1994)

A large range is also observed in the calculated initial εNd$_i$ values. The investigated LCT pegmatites exhibit a broad range for εNd$_i$ from 5.4 to −3.0 (Fig. 6.1a–c). A similar observation can be drawn for their depleted mantle model ages (TDM) that range between 2900 and 2400 Ma for the Bikita pegmatite field, 2800–2350 Ma for the Londonderry pegmatite field; 2500–2250 Ma for the Mount Deans pegmatite field; 2800 and 2550 Ma for the Cattlin Creek pegmatite, and 2950–2750 Ma for the Wodgina pegmatite. These TDM model ages roughly coincide with the crystallisation ages, providing an argument that the pegmatites crystallised almost directly from fractionated mantle melts.

In order to estimate the degree of REE fractionation, the fractionation factor f(Sm/Nd) is calculated which measures Sm/Nd enrichment in a given reservoir relative to CHUR (Chondritic Uniform Reservoir; DePaolo and Wasserburg 1976;

Roddaz et al. 2007). The calculated fractionation factor for the LCT pegmatites are compared to the geological background values for possible source rocks from the Zimbabwe, Yilgarn and Pilbara Craton (Fig. 6.1a–c). It is obvious that the samples that exhibit elevated $^{147}Sm/^{144}Nd$ ratios plot in the field of REE depletion. This further supports the suggestion that late stage replacement processes and related element redistribution affected the LCT pegmatites. However, the arguments of such late stage replacement processes and element redistribution hamper the identification of possible source rocks.

Most of the LCT pegmatite samples have Nd isotopic compositions close to depleted mantle, suggesting that they might have been derived directly from a depleted mantle source (Fig. 6.1d), with negligible Archean crustal contamination. But some of the samples have such low εNd_i values that indicate the involvement of early crust. This contrasting behaviour concerns also the samples from Bikita and may elucidate the need for additional Rb–Sr analyses as complementary to Sm/Nd isotope data.

6.2 Lithium Isotope Analysis on Selected Mineral Phases

Analyses of the Li isotope system are increasingly used to trace geological processes in LCT pegmatite systems and involve magmatic fractionation (Tomascak 2004; Maloney et al. 2008; Magna et al. 2010; Barnes et al. 2012) or hydrothermal alteration processes (Teng et al. 2006; Liu et al. 2010). The lithium isotope analysis takes benefits from the large relative mass difference between the two stable Li isotopes ^6Li and ^7Li of approximate 17 wt%. Thus, strong isotopic fractionation values up to δ^7Li at 60‰ can be observed (Hoefs 2018). In silicate melts, Li occurs in tetrahedral bonding with oxygen and due to its incompatible behaviour becomes enriched in the residual melt (Wenger and Armbruster 1991). The mass difference between the two Li isotopes influences their allocation into different mineral structures. The lighter ^6Li isotope is preferentially incorporated in more highly coordinated (octahedral) mineral structures (Wunder et al. 2007). Thus, the increasingly Li enriched residual melt becomes more ^7Li rich during magmatic fractionation.

Lithium abundance and isotope composition was determined for 49 samples of selected mineral phases (feldspar, quartz, mica, pollucite, petalite, garnet, beryl, tourmaline and spodumene) from LCT pegmatites. Also one pollucite sample from the Tanco LCT pegmatite deposit was analysed. The Li contents in the pegmatites range between 26 and ~32,500 ppm Li, reflecting the chemical composition of the minerals (Dittrich 2016). The δ^7Li values range between 0.06 and 31.92‰ (Table A22). The highest values at >13‰ are recorded in quartz and beryl. This is in good agreement with the general crystallisation sequence (Fig. 3.6) where late stage quartz and late stage beryl were formed from highly fractionated and thus more δ^7Li rich fluids. The measured mineral separates exhibit similar δ^7Li values when compared to other LCT pegmatite systems worldwide (Fig. 6.2). When compared to data from different granitic systems worldwide, the LCT pegmatites display higher δ^7Li values than most

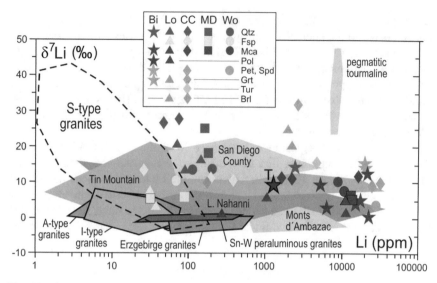

Fig. 6.2 Li contents and Li isotopic data from LCT pegmatite mineral separates compared to global pegmatites and granites. T Tanco, pollucite; Little Nahanni pegmatite group, NW Territories, Superior Province Craton, Barnes et al. (2012); Tin Mountain, Black Hills, South Dakota, USA, Teng et al. (2006); San Diego County tourmalines, California, Maloney et al. (2008); Monts d'Ambazac mica, French Massif Central, Deveaud et al. (2015); Edeko tourmalines, Nigeria, Abduriyim et al. (2006), and further tourmalines. Data for S-, I- and A-type granites: Bryant et al. (2004), Teng et al. (2004, 2006, 2009), Magna et al. (2010). Sn–W peraluminous granites, Gemeric Unit, Western Carpathians, Slovakia, Magna et al. (2010); Erzgebirge, Romer et al. (2014)

of the granites. As can be expected during magmatic fractionation, the incompatible Li becomes more enriched in the residual melt or fluid. Also this residual melt or fluid is enriched in δ^7Li. Although, the investigated samples yield distinct higher δ^7Li values than most of the A- and I-type granites (Teng et al. 2004, 2006, 2009; Magna et al. 2010; Bachmann et al. 2014) with only the S-type granites displaying a comparable broad δ^7Li range (-3 to 47 δ^7Li), no unequivocal relation of the LCT pegmatite systems to one specific A-, I- or S-type granite suite can be ascertained.

6.3 Fluid Inclusion and Pressure-Temperature Data

The pegmatites are hosted in Archean greenstone belts which consist mostly of greenschist and amphibolite facies metabasites. These rocks allow estimates on pressure and temperature (P–T) of the regional metamorphism. The mineral chemistry of green amphiboles in assemblages with plagioclase (albite and oligoclase), epidote, sphene, chlorite and quartz has been analysed by electron microprobe in zoning profiles. Pressure and temperature were estimated by applying the amphibole geothermobarometer of Zenk and Schulz (2004). At Cattlin Creek, the green Ca-amphiboles display prograde zonations with magnesio-hornblende in the cores and tschermakitic hornblende in the rims. The amphibole rims crystallised at a maximum of 620 °C/6.5 kbar. At Londonderry, the green Ca-amphiboles are markedly zoned with pale actinolite cores and magnesio-hornblende and ferro-hornblende rims. The cores crystallised at 400 °C/2 kbar, whereas the rims equilibrated at maximal 575 °C/5 kbar, corresponding to a prograde metamorphism. In amphibolites surrounding the Bikita pegmatites, one observes green Ca-amphiboles with actinolitic rims, and magnesio-hornblende and ferro-hornblende cores. They correspond to a retrograde P–T evolution from 575 °C/3.7 kbar to 450 °C/2 kbar. Although slight differences exist, the regional metamorphism in the greenstone belt evolved along typical thermal gradients in medium pressure–medium temperature orogenic belts.

6.3.1 Fluid Inclusion Studies

Fluid inclusions were encountered in almost all thin sections from the Bikita and Mount Tinstone pegmatite (Wodgina), and investigated during this study.

At Bikita the fluid inclusion study comprises quartz, pollucite, apatite and petalite. Highest homogenisation temperatures were observed for petalite from 370 to 400 °C. Quartz hosted inclusions homogenised to the liquid from 280 to 360 °C, those in pollucite from 280 to 350 °C. In contrast, apatite crystallised at relatively low temperatures from ~215 to 265 °C. The mineral forming fluid consists of $H_2O–CO_2–CH_4–NaCl_{eq}$ (Fig. 6.3a). Salinities in the inclusions in all minerals range from 2 to 15 wt% $NaCl_{eq}$. During quartz and pollucite crystallisation (280–360 °C) the fluid unmixed or "boiled" into a gas and aqueous phase. Dissolved elements in the fluid partitioned into the aqueous phase, whereby the salinity increased (Dittrich 2016; Richter et al. 2015b).

The minerals spodumene, quartz and apatite from the Mount Tinstone pegmatite at Wodgina exhibit trapping temperatures from >500 to 470–320 and 275–220 °C, respectively, at pressures between 2 and 3 kbar (Fig. 6.3b). The minerals crystallised from a homogenous $H_2O–CH_4–CaCl_2eq$ fluid with a salinity of about 26 wt% $CaCl_2eq$ by cooling, exsolution and mixing with formation water. The chemical composition of the fluid is controlled and dominated by Li, subordinate K, Ca, Na and Sn, Ta and Cs as trace components. According to mineral precipitation from 500 to 220 °C, the fluid chemistry changed towards higher contents of incompatible and mobile elements, such as Cs, P, F, Na and Ca. The results of the fluid inclusion petrography on the Bikita pegmatite field and Wodgina pegmatite reveal that the fluid inclusion assemblages at both localities contain a considerable amount of gaseous phases. Raman spectroscopy investigations demonstrate that the gaseous phase is made up of a mixture of CO_2 and CH_4 (Richter et al. 2015a, b).

The carbon isotope analyses on fluid inclusion gas was found to be useful for a further discrimination of the source of the fluids that are involved in the formation of the pegmatites. The fluids may have a mantle origin, their sources may be the dehydration of mafic volcanic rocks, or decarbonatisation of sedimentary and/or meta-

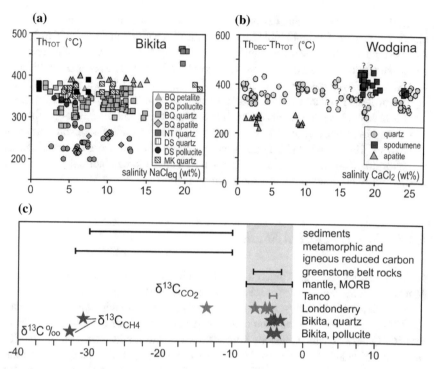

Fig. 6.3 Data from fluid inclusion studies in LCT pegmatites. **a** Microthermometry of fluid inclusions in minerals from Bikita pegmatites, in salinity (NaCl$_{eq}$) versus homogenisation temperatures. BQ—Bikita Main Quarry; NT—Nigel Tin; DS—Dam Site; MK—Mauve Kop. **b** Salinity (CsCl$_{2eq}$) versus homogenisation temperatures for the selected minerals in Wodgina pegmatites. **c** Comparison of $\delta^{13}C_{CO2}$ of sample gas in fluid inclusions from the Bikita and Londonderry pegmatites to data from Tanco pegmatite (Taylor and Friedrichsen 1983), mantle/MORB and greenstone belt rocks (Lüders et al. 2015), metamorphic and igneous reduced carbon (Des Marais 2001) and sediments (Carreira et al. 2010)

morphic rocks. The investigations were performed in cooperation with Dr. Volker Lüders at the German Research Centre for Geosciences (GFZ) at Potsdam. A total of 19 mineral separates (3–10 g) including quartz, pollucite and spodumene from the Bikita and Londonderry pegmatite fields, as well as the Wodgina pegmatite were handpicked and analysed. The stable C, and when present N, isotope ratios for CO_2 and CH_4 are analysed using a crushing device interfaced to an isotopic ratio mass spectrometer (IRMS). The full procedure of sample preparation and analytical conditions of this recently established method is described in Plessen and Lüders (2012) and Lueders et al. (2015). With the exception of two pollucite samples from the Bikita pegmatite field (BQ21, DS01), no N was detected within the investigated gases. Methane is present as traces with measurable amounts only present in two samples from the Bikita pegmatite field. The measured $\delta^{13}C$ in CH_4 values are:

$-30.8\permil$ for inclusions in quartz (WP50) and $-32.8\permil$ for inclusions in pollucite (DS01), respectively.

Although ascertained via freezing/heating experiments in all samples, CO_2 is encountered in measurable amounts only in six samples from the Bikita pegmatite field and four samples from the Londonderry pegmatite field (Fig. 6.3c). The obtained results reveal that the $\delta^{13}C$ in CO_2 values range between -3.2 and $-4.5\permil$ for the Bikita pegmatite field and between -4.8 and $-13.5\permil$ (average $-7.6\permil$) for the Londonderry pegmatite field. With the exception of sample L030 ($\delta^{13}C$ in $CO_2 = 13\permil$) from the Londonderry pegmatite field the $\delta^{13}C$ in CO_2 values cover a narrow range between -3.2 and $-6.7\permil$ (average $-4.1\permil$). This implies that the $\delta^{13}C$ in CO_2 values of the fluid inclusion gases are dominated by the supply of CO_2 during a major fluid event, and that the CO_2 rich fluid inclusions are probably representative for the discrimination of the origin of the fluids that were involved in pegmatite formation. The C isotopic compositions from the Bikita and Londonderry pegmatite fields plot within the field outlined by Lueders et al. (2015) to represent either mantle or MORB C isotopic compositions of -1.5 to $-8.0\permil$, or greenstone belt rocks with -3 to $-7.0\permil$ (Fig. 6.3c). The obtained $\delta^{13}C$ in CO_2 values are comparable with the -3.7 to $-4.7\permil$ $\delta^{13}C$ in CO_2 determined by Taylor and Friedrichsen (1983) on quartz and beryl from the Tanco pegmatite. The data suggest that the pegmatite-forming melt and associated fluids either derived from mantle material, with no or only limited signs of crustal contribution, or that fluid involved in the formation of the pegmatite intensely interacted with the surrounding greenstone belt lithology and inherited its C isotopic signature from the degassing greenstone belt rock.

References

Abduriyim A, Kitawaki H, Furuya M, Schwarz D (2006) "Paraíba"-Type Copper-Bearing Tourmaline from Brazil, Nigeria, and Mozambique: chemical fingerprinting by LA-ICP-MS. Gems and Gemology 42(1):4–21

Bachmann K, Seifert T, Magna T, Neßler J, Gutzmer J (2014) Li isotopes and geochemistry of Li–F–Sn greisen from the Zinnwald deposit, Germany. Goldschmidt Conference 2014, Sacramento, USA, abstracts volume, p 93

Barnes EM, Weis D, Groat LA (2012) Significant Li isotope fractionation in geochemically evolved rare element-bearing pegmatites from the Little Nahanni Pegmatite Group, NWT, Canada. Lithos 132–133:21–36

Bryant CJ, Chappell BW, Bennett VC, McCulloch MT (2004) Lithium isotopic compositions of the New England Batholith: correlations with inferred source rock compositions. Trans R Soc Edinburgh: Earth Sci 95:199–414

Carreira PM, Marques JM, Carvalho MR, Capasso G, Grassa F (2010) Mantle derived carbon in Hercynian granites. Stable isotopes signatures and C/He associations in the thermomineral waters, N-Portugal. J Volcanol Geotherm Res 189:49–56

DePaolo DJ (1981) Neodymium isotopes in the Colorado Front Range and crust–mantle evolution in the Proterozoic. Nature 291–5812:193–197

DePaolo DJ, Wasserburg GJ (1976) Inferences about magma sources and mantle structure from variations of $^{143}Nd/^{144}Nd$. Geophys Res Lett 3:743–746

Des Marais DJ (2001) Isotopic evolution of the biogeochemical carbon cycle during the Precambrian. Rev Mineral Geochem 43:555–578

Deveaud S, Millot R, Villaros A (2015) The genesis of LCT-type granitic pegmatites, as illustrated by lithium isotopes in micas. Chem Geol 411:97–111

Dittrich T (2016) Meso- to Neoarchean Lithium-Cesium-Tantalum- (LCT-) pegmatites (Western Australia, Zimbabwe) and a genetic model for the formation of massive pollucite mineralisations. Dissertation Faculty of Geosciences, Geoengineering and Mining, TU Freiberg/Saxony, Germany, 341 pp. http://nbn-resolving.de/urn:nbn:de:bsz:105-qucosa-228968

Hoefs J (2018) Stable isotope geochemistry, 8th edn. Springer, Berlin, p 437

Liu XM, Rudnick RL, Hier-Majunder S, Sirbescu MLC (2010) Processes controlling lithium isotopic distribution in contact aureoles: a case study of the Florence County pegmatites, Wisconsin. Geochem Geophys 11:1–21

Lüders V, Klemd R, Oberthür T, Plessen P (2015) Different carbon reservoirs of auriferous fluids in African Archean and Proterozoic gold deposits? Constraints from stable carbon isotopic compositions of quartz-hosted CO_2-rich fluid inclusions. Mineral Deposit 50:449–454

Magna T, Janoušek V, Kohút M, Oberli F, Wiechert U (2010) Fingerprinting sources of orogenic plutonic rocks from Variscan belt with lithium isotopes and possible link to subduction-related origin of some A-type granites. Chem Geol 274:94–107

Maloney JS, Nabelek PI, Sirbescu MLC, Halama R (2008) Lithium and its isotopes in tourmaline as indicators of the crystallization process in the San Diego County pegmatites, California, USA. Eur J Mineral 20:905–916

Milisenda CC, Liew TC, Hofmann AW, Köhler H (1994) Nd isotopic mapping of the Sri Lanka basement: update, and additional constraints from Sr isotopes. Precambr Res 66:95–110

Plessen B, Lüders V (2012) Simultaneous measurements of gas isotopic compositions of fluid inclusion gases (N_2, CH_4, CO_2) using continuous-flow isotope ratio mass spectrometry. Rapid Commun Mass Spectrom 26:1157–1161

Richter L, Seifert T, Dittrich T, Schulz B, Hagemann S, Banks D (2015a) Constraints on the magmatic-hydrothermal fluid. Evolution in LCT pegmatites from Mt. Tinstone, Wodgina Pegmatite District, North Pilbara Craton, Western Australia. Mineral resources in a sustainable world, 13th SGA Biennial Meeting 2015 Nancy, Proceedings vol 2, pp 529–532

Richter L, Lüders V, Hagemann SG, Seifert T, Dittrich T (2015b) Stable carbon isotopic composition of fluid inclusions from the Archean Bikita LCT pegmatite field. GeoBerlin 2015-Dynamic Earth from Alfred Wegener to today and beyond, 4–7 October 2015, GFZ German Research Centre for Geosciences, Berlin. GFZ Abstracts, pp 310–311. https://doi.org/10.2312/gfz.lis.2015.003

Roddaz M, Debat P, Nikema S (2007) Geochemistry of Upper Birimian sediments (major and trace elements and Nd–Sr isotopes) and implications for weathering and tectonic setting of the Late Paleoproterozoic crust. Precambr Res 159:197–211

Romer RL, Meixner A, Förster HJ (2014) Lithium and boron in late-orogenic granites—isotopic fingerprints for the source of crustal melts? Geochim Cosmochim Acta 131:98–114

Taylor BE, Friedrichsen H (1983) Light and stable isotope systematics of granitic pegmatites from North America and Norway. Chem Geol 41:127–167

Teng FZ, McDonough WF, Rudnick RL, Dalpé C, Tomascak PB, Chappell BW, Gao S (2004) Lithium isotopic composition and concentration of the upper continental crust. Geochim Cosmochim Acta 68:4167–4178

Teng FZ, McDonough WF, Rudnick RL, Walker RJ, Sirbescu MLC (2006) Lithium isotopic systematics of granites and pegmatites from the Black Hills, South Dakota. Am Mineral 91:1488–1498

Teng FZ, Rudnick RL, McDonough WF, Wu FY (2009) Lithium isotopic systematics of A-type granites and their mafic enclaves: Further constraints on the Li isotopic composition of the continental crust. Chem Geol 262:370–379

Tomascak PB (2004) Developments in the understanding and application of Lithium isotopes in the earth and planetary sciences. Rev Mineral Geochem 55:153–195

Wenger M, Armbruster T (1991) Crystal chemistry of lithium: oxygen coordination and bonding. Eur J Mineral 3:387–399

Wunder B, Meixner A, Romer RL, Feenstra A, Schettler G, Heinrich W (2007) Lithium isotope fractionation between Li-bearing staurolite, Li mica and aqueous fluids; an experimental study. Chem Geol 238:277–290

Zenk M, Schulz B (2004) Zoned Ca-amphiboles and related P–T evolution in metabasites from the classical Barrovian metamorphic zones in Scotland. Mineral Mag 68:769–786

Chapter 7
Genesis of Massive Pollucite Mineralisation in Archean LCT Pegmatites

7.1 Pegmatite Ages and Potential Source Granites

The U–Pb dating of the LCT pegmatites from the Yilgarn and Zimbabwe Craton yielded comparable Neo-Archean ages at 2650–2600 Ma. These ages coincide to data from the Tanco pegmatite in Canada. The 2650–2600 Ma LCT pegmatite formation within the Yilgarn, Zimbabwe and Superior cratons define the first major LCT pegmatite formation event. This event concerns a narrow time span which suggests that similar geological processes were active at various locations worldwide. In contrast, the Meso-Archean age of the Wodgina pegmatite on the Pilbara Craton coincides with an earlier and minor LCT pegmatite formation event (Figs. 1.4 and 7.1).

In the Yilgarn Craton, the formation of low-Ca granitoid melts is constrained to be between 2655 and 2630 Ma (Goscombe et al. 2009), but also the intrusion of a suite of mantle derived adakitic granitoid rocks (Martin et al. 2005) which coincided with most of the ages of the Londonderry, Mount Deans, and the Cattlin Creek pegmatites. Also, the emplacement of the low-Ca granitoid melts and adakites was accompanied by localised hydrothermal events that lead to the formation of major Au deposits within the Eastern Yilgarn Craton (Mueller et al. 2016). This episode was characterised also by a long period (2640–2615 Ma) of intense, but localised hydrothermal activity at elevated temperatures of 250–350 °C. The heat for the fluid flow was provided by the low-Ca magmatism (Goscombe et al. 2009). This period of combined magmatic and hydrothermal activity at the end of the Neo-Archean coincided well with the formation age of the LCT pegmatites and implies a close genetic relationships (Fig. 2.7). A derivation of the pegmatite melts from an adakitic source is supported by the whole rock Sm-Nd isotopes and the C stable isotope compositions of gas in fluid inclusions. These suggest a non-crustal source for the pegmatites, a depleted mantle, and mantle derived rocks as MORB or metabasalts.

© The Author(s), under exclusive license to Springer Nature Switzerland AG 2019
T. Dittrich et al., *Archean Rare-Metal Pegmatites in Zimbabwe and Western Australia*,
SpringerBriefs in World Mineral Deposits, https://doi.org/10.1007/978-3-030-10943-1_7

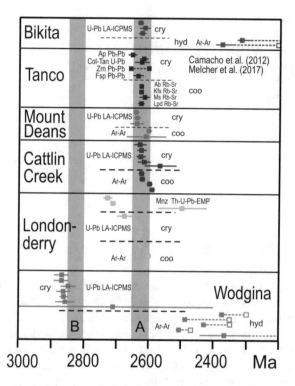

Fig. 7.1 Geochronological data from Archean LCT pegmatites (this study) and Tanco (Camacho et al. 2012; Melcher et al. 2017). Interpretation as crystallisation ages (cry), cooling age (coo) and age of hydrothermal/metamorphic overprint (hyd). A major, B minor LCT pegmatite crystallisation events after Tkachev (2011)

The development of the Neo- and Meso-Archean LCT pegmatites in the Zimbabwe, Yilgarn and Pilbara cratons was probably not a single stage process, as indicated by some of the Ar–Ar mica ages (Fig. 7.1). At least two stages of evolution, the early magmatic crystallisation and the later hydrothermal process can be outlined. The hydrothermal process can be related to the magmatic cooling (Yilgarn Craton), but also to a later overprint (Bikita, Wodgina).

The Neo- and Meso-Archean LCT pegmatites are commonly hosted in greenstone belt lithologies. Even when the pegmatite crystallisation ages allow to identify adjacent granitoid suites within the same age span, no direct field evidence for a connection to potential source granites could be observed. Contact relationships between the pegmatites and the surrounding greenstone belt lithologies explicitly display intrusive character. This suggests a remote source for the pegmatite forming melts. Although a source for the pegmatite forming melts from a deeper lithospheric level cannot be excluded, a thorough examination of the regional geological data allows the identification of possible source granitoids.

The Bikita pegmatites display the classical elemental and mineralogical zonation as established by Trueman and Černý (1982). The most evolved portions of the Bikita LCT pegmatite field are represented by the pegmatite ore body with the massive pollucite mineralisation. This mineralogical and geochemical trend unequivocally displays a NE to SW directed fractionation and suggests a probable source granite towards the NE of the Masvingo greenstone belt (Fig. 2.2a). Indeed, this region exhibits a low vertical gravity anomaly that was interpreted by Gwavava and Ranganai (2009) as evidence for granitoid intrusive bodies in the vicinity. A main structure within the Bikita pegmatite district is the NW–SE striking Gono fault which exhibits dominant normal and minor dextral movement. A second conjugated N–S striking system of faults points towards the location of the proposed source granite and may have served as conduits for the emplacement of the LCT pegmatites.

The emplacement ages of about 2870 Ma of the LCT pegmatites at Wodgina in the Pilbara Craton are coinciding with the magmatic intrusions of the Split Rock Supersuite (2890–2830 Ma). Although, no direct field evidence exists, the spatial and age relations (Figs. 2.15 and 2.16) suggest a strong genetic relationship of the pegmatites with the Numbana monzogranite as a member of the Split Rock Supersuite to the SE of the greenstone belt (Sweetapple and Collins 2002).

No field evidences for possible source granites are observed at the Londonderry pegmatite field in the Yilgarn Craton (Dittrich 2016; Dittrich and Seifert 2013). The general zonation with more evolved pegmatites in the northern part (Lepidolite Hill, Tantalite Hill) suggests a possible source granite towards the south. The area S of the Kangaroo Hill greenstone belt is dominated by granite gneiss domains and outcrops of several different types of granitic intrusions (Fig. 2.8). As most of the area is covered by post-Archean sedimentary rocks (Fig. 2.9a), no direct evidence for source granites can be established. A similar observation is made at Mount Deans. There, the higher specialised pegmatites are concentrated towards the central southern portion of the pegmatite field (Figs. 2.8 and 2.11). This indicates that the probable source granites may be located towards the N or NE. Outcrops of monzogranite and granodiorite with matching ages (Figs. 2.6 and 2.13) can be observed about 20–30 km SE of the Cattlin Creek pegmatite. However, as the structural geology of the Ravensthorpe Terrane is complex, no further predictions can be drawn for possible source granites.

7.2 Mineralogical and Geochemical Characteristics of LCT Pegmatites

A preliminary working model for the formation of massive pollucite mineralisation at Bikita was based on the idea that pollucite will start to crystallise from the highest evolved pegmatite melt contemporaneous with the remaining minerals. This should lead to a continuous increase of pollucite towards the massive mineralisation. However, no pollucite has been detected near the mineralisation. The general crystallisation sequence comprises three main stages: (I) early, (II) main, and (III) late crystallisation (Fig. 3.6). The early stage (I) is characterised by subsolvus crystallisation from a hydrous melt that forms the predominant portions of the plagioclase and K-feldspar contemporaneous with the crystallisation of petalite, spodumene, cassiterite and Nb–Ta oxide minerals. The main stage (II) is accompanied by ongoing fractionation of the melt and the exsolution of a fluid phase. It is characterised by the growth of most of muscovite and quartz as well as the occurrence of tourmaline, apatite and beryl. Muscovite is the predominant mica at the beginning of the main stage mineralisation and is followed later by more abundant lepidolite. Crystallisation of pollucite started at the end of the main stage and continued in the late stage. The late stage (III) crystallisation is dominated by mineral replacement reactions. The mineralogical changes within this stage are attributed to deuteric hydrothermal fluids that circulate through a consolidated, but still not fully cooled pegmatite system.

It is demonstrated that the element variation in the LCT pegmatites are controlled by their mineralogy and mode, e.g., K, Rb are enriched in domains with mica, Na is enriched in domains with dominating feldspar, and Sn is concentrated in samples with cassiterite (Fig. 3.1). Accordingly, the maximum Cs content that is feasible to accumulate without the presence of pollucite is detected in the mica bearing mineral assemblages that can contain about 1–2 wt% Cs_2O (Figs. 4.1 and 4.2). This is in good agreement with data reported from rare element pegmatites (Černý et al. 1985). Also higher concentrations of Cs, Rb, Li, K, Ta, Nb, Sn and F, as detected in the highest fractionated samples, are characteristic features for the geochemical development of LCT pegmatites (Trueman and Černý 1982). Although the massive pollucite bearing mineral assemblage from Bikita exhibits a similar behaviour and trends when compared to the other mineral assemblages, they do not correspond to the highest fractionated portions of the pegmatite. This leads to the suggestion that the massive pollucite mineralisation was formed prior to the highest fractionated portions with lepidolite dominated mineral assemblages. However, the most distinguishing feature of the massive pollucite bearing mineral assemblages is their extreme enrichment in Cs of up to 25 wt% Cs_2O, and the absence of intermediate Cs contents. This is interpreted to indicate that the formation of the massive pollucite mineralisation is not an essential part of the general fractional crystallisation of LCT pegmatites. Additional processes should have formed such a specific mineralisation at Bikita, as is discussed in more detail below.

The REE of the LCT pegmatites exhibit similar chondrite normalised distribution patterns (Fig. 4.4). The REE ratios range between 0.1 and 100 and are therefore comparable to most granite and LCT pegmatites (Černý 1982; René 2012). Except for Bikita, two different types of REE distribution pattern are present: the type-I REE pattern exhibits higher chondrite normalised REE ratios between ~10 and 100, shows a slight enrichment of LREE over HREE, and is often characterised by a prominent negative Eu anomaly. An exception are the type-I REE distribution pattern from the Wodgina pegmatites that yield a prominent positive Eu anomaly. The type-II REE pattern was only observed from pegmatites of Western Australia. Three samples from the Mount Deans pegmatite field exhibit a type-III REE distribution pattern (Fig. 4.4d) with similar REE ratios as in type-II (0.1–10) but with a strong depletion of the LREE compared to the HREE. REE bearing minerals as xenotime, monazite, zircon are only very rarely observed. This leads to the suggestion that most of the REE content is controlled by the overall presence of feldspars and more abundant other accessory minerals such as apatite and garnet. The predominance of feldspar explains the prominent negative Eu anomalies. According to Fowler and Doig (1983) almost the complete Eu^{2+} substitutes into feldspar relatively early during the magmatic differentiation. In contrast, positive Eu anomalies were observed from REE type-I samples from Wodgina (Fig. 4.4e). Fowler and Doig (1983) explained this by late stage and most probably external hydrothermal fluids that caused albitisation of a more calcic plagioclase under oxidising conditions and that preferential enriched the fluids in Eu^{3+}. From the Erzgebirge (Germany), less REE enriched pattern were described as typical for phosphorous-rich S-type granites and the more REE enriched type-II pattern are characteristic for P depleted A-type granites (René 2012; Seifert 2008). Alternatively the differences between type-I and type-II might be explained by the buffering behaviour of P (London et al. 1999, 1993; Colombo et al. 2012). Phosphorous can substitute for Al in alkali feldspars, but will be released from the silicates during late stage hydrothermal processes. Thus, during feldspar replacement by late stage hydrothermal processes beside P also the REE will be released (Vasyukova and Williams-Jones 2014). The REE and P then may have formed late stage apatite. Hence, type-II REE pattern can develop from primary and magmatic fractionation, and type-I REE pattern by the circulation of late stage hydrothermal fluids.

Fluid inclusions of selected mineral phases from the Bikita and Wodgina LCT pegmatites recorded comparable homogenisation temperatures which range from approximately 200–450 °C for Bikita and 200–500 °C for the Wodgina (Fig. 6.3a, b). Quartz and the Li minerals petalite (in Bikita) and spodumene (Wodgina) yield higher temperatures (300–450 °C). Pollucite (Bikita) and apatite (Bikita and Wodgina) exhibit lower entrapment temperatures (200–300 °C). This is in good agreement with the general crystallisation sequence, with most of the quartz, petalite and spodumene representing early stage minerals, whereas apatite and the pollucite from the massive pollucite mineralisation were formed during the main to late stage of the crystallisation. The $\delta^{13}C_{CO2}$ values of fluid inclusions point to a mafic or ultramafic source, either directly from the mantle or MORB, or alternatively from local remobilisation of the surrounding greenstone belt lithologies (Fig. 6.3c; Richter et al. 2015a, b).

Whole rock Sm/Nd analysis reveal that some samples yield extreme high $^{147}Sm/^{144}Nd$ ratios (>0.3) indicating extreme fractionation of the REE. This may be related to the late stage replacement processes and element redistribution by replacements of feldspar and primary Nb-, Ta- and Sn-oxides. Comparison of the fractionation factor ($f_{(Sm/Nd)}$) in LCT pegmatites to data from various rock suites of the Zimbabwe, Yilgarn and Pilbara cratons exhibit comparable values which differ from the Archean crust (McLennan et al. 1993). The pegmatite have Nd isotopic compositions close to depleted mantle, suggesting that they may have been derived directly from a depleted mantle source with only minor Archean crustal contamination (Fig. 6.1d). These values may reflect the original composition of the source, or are product of the extreme fractionation processes during pegmatite genesis.

The Neo- and Meso-Archean LCT pegmatites have similar δ^7Li values as other LCT pegmatite systems worldwide (Fig. 6.2). Only a few quartz and the beryl samples exhibit higher δ^7Li values when compared to the global pegmatites and granites. This again may be explained by late and thus more ^{7}Li rich fluids that were present during the crystallisation of those minerals. Comparison to potential source granite data allowed no correlation to a specific type of granite.

The most characteristic feature of the massive pollucite mineralisation is the ability of pollucite to incorporate tremendous amounts of up to, theoretically, 43 wt% Cs_2O. An enrichment factor of 215,000 is needed to concentrate Cs from average crustal levels to end member pollucite concentrations. Apart from pollucite (65 wt%), only mica (20 wt%), quartz (7 wt%) and feldspar (6 wt%) are present in the massive pollucite mineralisation at Bikita and Tanco (Stilling et al. 2006). Another characteristic of this mineral zone is the apparent lack of high density minerals such as cassiterite, zircon, columbite, and microlite group minerals. Small traces of Sb-bearing tantalite group minerals (stibiotantalite) at Bikita and wodginite group minerals at Tanco are exceptions. As the pollucite is concentrated in the massive mineralisation, there appears a large compositional gap, when geochemical data from the various pegmatite mineral zones is plotted versus Cs (Fig. 7.2). For many elements, such as Al_2O_3, Rb and F, an enrichment of Cs along a magmatic fractionation trend can be observed. In the muscovite- and lepidolite-dominated mineral zones which represent the late stage crystallisation, a maximal enrichment of Cs to 1–2 wt% can be observed. This is in marked contrast to the pollucite-bearing zones with 20–35 wt% Cs at Bikita and Tanco. Following these observations, one can outline two stages of Cs enrichment. The early stage enrichment follows a common magmatic fractionation. The second stage is decoupled from the magmatic fractionation and is characterised by a discontinuous strong Cs enrichment of more than one order of magnitude, reaching the compositions of Cs-analcime and Na-pollucite (Fig. 7.2).

The internal composition of the pollucite is very inhomogeneous at Bikita. In BSE images (Fig. 3.2d–f) the patchy distribution of dark coloured pollucite with about 20 wt% Cs_2O and corresponding to Cs-analcime (Pol_5–Pol_{50}), is crosscut by several fingers of bright coloured pollucite with 20–30 wt% Cs_2O, which is a composition of Na-pollucite (Pol_{50}–Pol_{95}). The Cs-analcime and Na-pollucite matrix is also invaded by a vein like network of a late generation of pollucite that is characterised by further enrichment of Cs, up to 37 wt% Cs_2O, but still Na-pollucite. End member pollucite Pol_{95}–Pol_{100} with 42.94–45.16 wt% Cs_2O was never detected. In all studied pollucite samples, there are portions with Cs-analcime compositions which occur as lamellar bands within brighter, thus more Cs-enriched Na-pollucite. Teertstra and Černý (1995) interpreted such lamellae as the result of exsolution of Cs–Al-enriched and Na–Si-enriched pollucite during cooling of the primary pollucite.

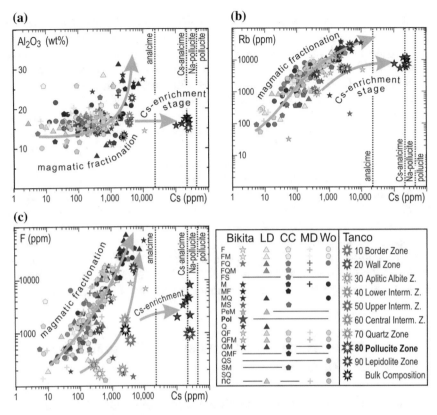

Fig. 7.2 a–c Comparison of selected elements and fractionation indicators in pegmatites from Bikita and Western Australia (this study), and Tanco (Stilling et al. 2006). Arrows indicate Cs enrichment by magmatic fractionation and a later separate stage. Dotted lines indicate the Cs contents of analcime, Cs-analcime, Na-pollucite and pollucite

7.3   Late Stage Hydrothermal Processes

At Bikita, the vein-like pollucite which is crosscutting the earlier pollucite generation is observed to merge into a lepidolite dominated vein network (Fig. 3.2). A shift from Cs-analcime towards Na-pollucite composition in close spatial association with the lepidolite filled veins is attributed to the activities of late stage hydrothermal fluids. Although only limited material was available, similar textural observations are made for the investigated pollucite sample from the Londonderry pegmatite field and from the Tanco LCT pegmatite deposit. The late hydrothermal event was accompanied by a significant redistribution of Cs. Further prominent late stage replacement phenomena are found. Mica and quartz replace K-feldspar. This replacement predominantly started along cleavage plains, small cracks and in interstices as small inclusions of sericitic micas and proceeded to larger grains of lepidolite and muscovite (Fig. A7g).

At Cattlin Creek, the K-feldspar is almost completely replaced by cleavelandite, purple lepidolite and quartz. Late stage replacement also affects other minerals such as mica, apatite, beryl, tourmaline, and the Nb-, Ta- and Sn-oxide minerals. Detailed combined SEM-EDX investigations could demonstrate that the rims of micas can be enriched in Cs (Fig. A10g–i). Most of the rims contain only <0.5 wt% Cs, but certain minerals exhibit rims that contain several percent of Cs. The highest Cs contents (up to 20 wt%) were found in lepidolite-lamellae from Bikita. Also considerable Cs enrichment was documented for rims of tourmaline and for rims and along cracks in beryl (Fig. 3.5f, g, i). Such late stage hydrothermal replacement processes are observed within the entire pegmatite field. This suggests the existence of a prominent hydrothermal fluid event that interacted with the melt and/or the already crystallised portions of the pegmatite during or shortly after its consolidation. At Bikita and Wodgina, the $^{40}Ar-^{39}Ar$ cooling ages of mica are considerably younger (>200 Ma) than the pegmatite crystallisation (Fig. 7.1). This indicates that both LCT pegmatite systems were subjected to a later thermal event that was not related to the initial crystallisation. In the Yilgarn Craton, the $^{40}Ar-^{39}Ar$ cooling ages of mica coincide with the pegmatite crystallisation ages (Fig. 7.1). Therefore these pegmatites experienced no major hydrothermal overprint late after their crystallisation.

7.4   Structural Setting of Massive Pollucite Mineralisation in LCT Pegmatites

The key features of massive pollucite mineralisation that separate those from other LCT pegmatite systems can be summarised emphasising the following essential characteristics: (a) the structural position in the upper portions of the largest sheets in a LCT pegmatite field; (b) extreme enrichment in Cs when compared to the hosting LCT pegmatite; (c) characteristic pollucite typical mineralogical and mineral-chemical compositions, and (d) evidences for extensive late stage hydrothermal processes.

At Bikita, massive mineralisation of pollucite is observed only in two locations (Dittrich 2016; Dittrich et al. 2015). They occur as irregular lenticular ore bodies that exhibit sharp contacts with the surrounding pegmatite zones. The massive pollucite lens is located between the border zone and the intermediate zones of the pegmatite sequence (Figs. 2.3 and 2.4). At the Bikita Main quarry pegmatite sheet which is 30–40 m in total thickness and displays a dip of 30–45°, the massive pollucite mineralisation is up to at least 100 m long and up to 15 m thick. At the Bikita Dam Site locality, the massive pollucite mineralisation forms a small lens of about 2 m in thickness. Similar observations were reported by Stilling et al. (2006) and Crouse and Černý (1972) for the massive pollucite mineralisation within the Tanco LCT pegmatite deposit (Fig. 7.3). At this point it has to be noted that the cross sections through the Tanco pegmatite show different interpretations for the zonation. In London (2008, Fig. 7-3), the cross section was redrafted from Stilling et al. (2006),

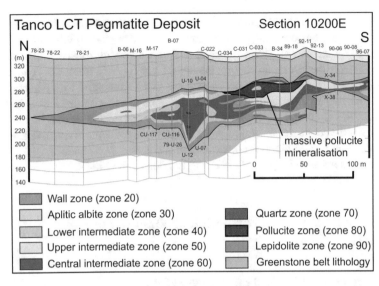

Fig. 7.3 Section 10200E of the Tanco LCT pegmatite, modified after Stilling et al. (2006)

and the quartz zone (zone 70) and the pollucite zone (zone 80) were mistakenly interchanged. All other available cross sections within the other reports (Crouse and Černý 1972; Simpson 1974) display the correct illustration of the pegmatite mineral zones.

Another similarity of the Bikita and Tanco LCT pegmatite deposits is that the largest bodies of the massive pollucite mineralisation are found in the main sheets of the pegmatite systems. When the Bikita Main Quarry pegmatite and the Dam Site location are considered (Fig. 2.4), a relation between the size of the massive pollucite mineralisation and the total thickness of the pegmatite sheet can be stated. Ferreira (1984) documented subordinate massive pollucite mineralisation in separate sheets directly underlying the Tanco pegmatite. Also Černý and Harris (1973) reported massive pollucite mineralisation from the Buck Claim and Odd West pegmatites further 10 km E of the Tanco pegmatite. Such occurrences are indicative for the presence of larger ore bodies in larger pegmatite sheets. Furthermore it can be deduced, that conditions of pollucite formation were active within the entire pegmatite field and were not only restricted to the highest evolved portions. The location of small massive pollucite bodies can be used as an exploration tool for likely larger bodies in genetically related larger pegmatite sheets close by.

The interpretation that the favourable conditions for pollucite formation could be expanded to the entire pegmatite field is in contrast to the general concepts of pollucite formation (Černý 1982; London et al. 1998, 2017). There, the pollucite formation is considered as a very late stage in the development of pegmatites. Pollucite should crystallise either directly from a highly fractionated granitic melt (Sebastian and Lagache 1990; London et al. 1998), or via precipitation from a hydrothermal fluid (Redkin and Hemley 2000; Endo et al. 2013; Yokomori et al. 2014). Also, in these concepts, pollucite crystallisation is interpreted to be restricted to the highest evolved portions within the core zone of the pegmatite. However, the observations at Bikita and Tanco signal that formation and crystallisation history of massive pollucite differ markedly from the classical zonation model by Černý (1982), Simmons et al. (2003), and London (2008). The extreme enrichment of Cs compared to other pegmatite zones, the lack of intermediate Cs concentrations in other mineral zones, the Cs gap, and the structural position of massive pollucite mineralisation in the upper portions of the pegmatite sheets are in conflict with the concepts of ongoing fractional crystallisation and continuous enrichment of incompatible elements in the residual melt/fluid as outlined in Černý (1982) and London et al. (1998).

7.5 Concepts for the Formation of Massive Pollucite Mineralisation

7.5.1 Melt Immiscibility and Separation

Comparison of the Cs contents and mineral modes in the pegmatite mineral zones demonstrates a mineralogical control (Fig. 7.2). The maximal Cs content that is observed without the presence of pollucite is about 1–2 wt% of Cs_2O. If the crystallisation of pollucite should be explained by ongoing fractional crystallisation with an associated gradual increasing content of Cs, this should be obvious by a continuous transgression from about 2 wt% Cs_2O to about 23 wt% Cs_2O, and a gradually increasing content of pollucite or other Cs-bearing minerals within the pegmatite mineral zones. Such continuous enrichment trends are observed for Na, K, Rb, Li, Tl, but not for Cs (Figs. 4.1, 4.2 and 7.2b). On one hand there is a considerable gap in the Cs contents between the massive pollucite mineralisation and the other mineral zones (Fig. 4.1). On the other hand the structural and mineralogical observations at Bikita indicate that the massive pollucite mineralisation is associated with the main stage crystallisation and does not relate to late pegmatite mineral zones. As a consequence, the massive pollucite mineralisation does not represent the late stages of a continuous fractionation and enrichment process. This allows the conclusion that the formation of massive pollucite mineralisation involves two stages. The first stage encloses the classical fractional crystallisation with enrichment of incompatible elements. The second stage concerns the extreme enrichment of Cs, which appears distinctly separated from the general development of LCT pegmatites.

As demonstrated by the geochemical data (Figs. 4.1 and 7.2), the maximum content that is found to be accumulated during fractional crystallisation is about 1–2 wt% Cs_2O, which allows the crystallisation of Cs-analcime. Experimental studies by Liou (1971) demonstrated that analcime formation contemporaneous with albite and quartz can commence at high temperatures (>500 °C) and high fluid pressures (>500 bar) from high alkaline melts (pH >11; Inagaki et al. 2006). Experimental studies by Peters et al. (1966) and Roux and Hamilton (1976) defined a restricted pressure (5–13 kbar) and temperature (600–640 °C) space in which direct crystalli-sation of analcime from a water rich silicate melt is possible (Fig. 7.4). Additional components like K (Peters et al. 1966), or F and Cl (Roux and Hamilton 1976) will increase the thermal stability of analcime and decrease the solidus. The defined P–T space in the study by Peters et al. (1966) is comparable to the experimental results of London (1984) in the system $LiAlSiO_4$–SiO_2–H_2O, and can explain the observed paragenesis of pollucite with spodumene or petalite. However, although not impos-sible, it appears unlikely that a LCT pegmatite melt will pass through the narrow P–T field for analcime crystallisation as defined by Peters et al. (1966) and Roux and Hamilton (1976). Certainly, such restricted P–T conditions could explain the extreme rarity of pollucite in LCT pegmatites. Therefore, it is proposed that at a relatively early stage of pegmatite formation, the pegmatite melt passed through this critical P–T space, which then enabled the crystallisation of analcime. As an early fractional crystallisation of the melt was already able to enrich incompatible Cs, a Cs-analcime composition (>2.1 wt% Cs_2O) can be expected. This Cs-analcime will then form small inclusions or interstitial grains in and between other phases as quartz, feldspar, mica and beryl. Such textural positions give evidence that Cs-analcime should have crystallised prior to the final consolidation of the silicate melt. Subsequent substitu-tion of Na by Cs within the analcime structure (Lagache et al. 1995) then would be capable to transform Cs-analcime to Na-pollucite and towards an end member pollu-cite. This could explain the occurence of relatively small mass of pollucite discovered within the drill hole from the Londonderry pegmatite field, and the location of tiny Cs rich mineral inclusions that resample pollucite compositions. Furthermore, similar occurences of pollucite are reported from a number of other pegmatites worldwide. A primary magmatic analcime crystallisation implies that such melts originated from depths of 20–50 km (Woolley and Symes 1976), or >35 km (Pearce 1970; Ferguson and Edgar 1978). Such crustal depths also correspond with the generation of adakite melts as proposed by Mueller et al. (2016).

Primary magmatic crystallisation of Cs-analcime and its substitution towards Na-pollucite alone can not explain the structural position of the massive pollucite min-eralisation at the Bikita and Tanco pegmatites. A process for accumulation of larger amounts of melt, compatible with Cs-analcime compositions in the upper portion of the pegmatite sheet prior to the crystallisation of the Cs-analcime, and the remain-ing portions of the pegmatite appears necessary. Immiscibility could be capable to produce sufficient amounts of melt with analcime composition. Thomas et al. (2011) and Vasyukova and Williams-Jones (2014) demonstrated that melt fluid immiscibility can effectively concentrate REE and Be in granitic pegmatite systems. Furthermore, Thomas and Davidson (2012) proposed that besides the two immiscible melts, an additional liquid phase is involved in the pegmatite formation and that separation into these phases occurs relatively early during pegmatite crystallisation. This liq-uid melt immiscibility is capable to enrich Cs, but the degree of enrichment was not determined in full detail. However, Thomas et al. (2011) stated that melt-melt immis-cibility is capable to enrich Be from about 10 ppm from a host granite to 10,000 ppm within melt inclusions in pegmatite, corresponding to an enrichment factor of 1000.

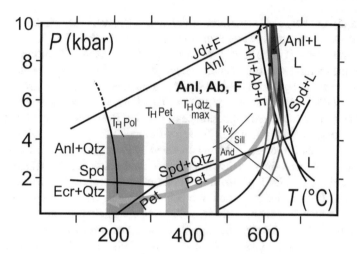

Fig. 7.4 P–T diagram illustrating the phase stability fields and reactions in the system NaAlSiO$_4$–SiO$_2$–H$_2$O, considered as applicable for pegmatite crystallisation at Bikita. Albite (Ab), analcime (Anl), eucryptite (Ecr), fluid (F), jadeite (Jd), melt (L), petalite (Pet), quartz (Qtz), spodumene (Spd) involving reactions (black lines), stability fields and granitoid solidus according to experimental data in Liou (1971) and London (1984). Conditions for the melt with analcime (Anl+L) after Roux and Hamilton (1976) in blue lines. Coloured bars indicate the ranges for the homogenisation temperatures (T$_H$) of the fluid inclusion assemblages in pollucite (Pol), petalite and quartz at Bikita. Arrow indicates the proposed evolutionary P–T trend for the Bikita LCT pegmatite deposit

Liquid melt immiscibility appears as capable to separate larger amounts of melt with analcime composition. Filter pressing or filter differentiation (Propach 1976) is an alternative process. During this process crystals are separated from melts by sieving. Filter differentiation is reported from various geological settings including melt segregation from its initial melting region, for mafic and felsic magmatites, and pegmatites (Lux et al. 2007; Grosse et al. 2010). More recent interpretations of filter pressing include the exsolution of a gaseous phase next to the new melt and residual crystals (Sisson and Bacon 1999; Pistone et al. 2015). They explain the separation of the different phases due to build-up of pressure (second boiling event) which segregate highly evolved melts and fluids from crystal mush.

Regardless of the process, two circumstances will enhance the Cs-analcime melt separation. Fluxing components (London 1987) commonly are interpreted to reduce the crystallisation temperatures that will cause undercooling and thus retard the crystallisation of Cs-analcime. Apart from the contribution of fluxing components, the density of the Cs-analcime melt will be comparably low. The theoretical density of the pollucite-analcime solid solution series ranges between 2.21 and 2.94 g/cm^3 for end members analcime (Pol$_{10}$) and pollucite (Pol$_{100}$; Table 1.1). A Cs-analcime melt will have even a lower density than the remaining pegmatite melt that is predominantly composed of a mixture of feldspars, quartz and micas with densities between 2.54 and 3.30 g/cm^3. Consequently, this relative density contrast will induce a gravity driven separation, an ascent, and an accumulation of the Cs-analcime melt in the upper portions of the pegmatite sheet. The gravity driven separation further explains also the lack of higher density minerals of early crystallisation stages such as cassiterite, zircon, columbite or microlite group minerals. An emplacement of a separated melt into an already partly consolidated pegmatite sheet is supported by the observation of irregular fragmentation and reworking at the Bikita Main Quarry. Based on the data set of Stilling et al. (2006) for the zonal and bulk compositions of the Tanco pegmatite, only 0.5–1 vol% of the total volume would be sufficient to form the massive pollucite mineralisation.

7.5.2 *Extreme Cesium Enrichment in Massive Pollucite Mineralisation*

Without the magmatic fractionation of analcime, only about 1–2 wt% Cs_2O are observed in the LCT pegmatites. Compositions of fertile granites represent the starting composition of the pegmatite melt and can have enrichment of up to 10 times of the average crustal composition in Be, Cs, Rb, Li, Nb, Ta and Sn (Taylor and McLennan 1985; Selway et al. 2005). Under the premise that during the process of fluid melt immiscibility the Cs will be similarly enriched as documented for Be by Thomas et al. (2011), then a partial melt potentially will contain 40,000–50,000 ppm Cs. This is equivalent to 4.24–5.3 wt% of Cs_2O. Such a Cs-analcime melt has a maximum density range of 2.25–2.58 g/cm^3, still allowing gravitational separation and emplacement in the upper portion of the pegmatite sheet. Although the experimental study of London et al. (1998) did not include immiscibility processes, it could demonstrate that it is possible to crystallise Cs-analcime (Pol_5–Pol_{50}) from a silicic artificially Cs enriched melt with about 5 wt% Cs_2O. As larger amounts of Cs can not enter quartz, feldspars, and micas, the excess Cs will be enriched into the remaining melt and fluids, when these minerals start to crystallise (Redkin and Hemley 2000). This process accompanies the entire crystallisation of the pegmatite melt, including the time span of analcime melt separation. Particularly the fluid phase is then capable to react with the immiscible analcime melt, leading to its enrichment. When the crystallisation of the separated analcime melt starts, the Cs is incorporated into the crystal lattices of any phase along the analcime–pollucite solid solution series, as Cs-analcime or even Na-pollucite (Lagache et al. 1995). This process is unambiguously related to the magmatic stage of pegmatite consolidation. The observations of Na and Cs enriched lamellae suggest that subsolidus exsolution of the pollucite takes place during consolidation of the pollucite.

7.5.3 The Role of Feldspars for Cesium Enrichment

K-feldspar and plagioclase form vast portions of LCT pegmatite systems. The behaviour of feldspars during cooling of the pegmatite is, therefore, interpreted to play a major role for the Cs enrichment in late stage hydrothermal fluids. In highly evolved pegmatites feldspars may yield 300–500 ppm Cs. If these feldspars are replaced by micas, as observed in LCT pegmatites, some of the Cs from the feldspars is remobilised. Furthermore, during cooling, the feldspars exhibit complex subsolidus processes that are concomitant with a re-arrangement of the crystal lattice and its chemical composition. The major processes are exsolution, recrystallisation, development of twins and mineral self ordering processes (Yund 1983; Sanchez-Munoz et al. 2012). Under temperature conditions of <300 °C, the feldspar lattice attempts to change to a higher symmetry and ordering (Ribbe 1983). Thus, it is suggested that during this process the elements which entered the crystal at high temperatures, but are not essential for the structure, like Rb and Cs, are removed and will be concentrated into the hydrothermal fluids.

Experiments by Ellis and Mahon (1977) could show that transport of Li, Rb and Cs by a hydrothermal fluid is more favourable at moderate temperatures from 200 to 300 °C than at higher temperatures (500–600 °C). After Cs was effectively removed from the crystal structure of the feldspars, Cs behaves highly incompatible and will remain within the late stage hydrothermal fluid. Cesium then is transported along interstitials or fractures to the parts of the pegmatite where Cs-analcime and/or Na-pollucite are concentrated to form the later massive pollucite mineralisation (Lagache et al. 1995). Even at this stage it is suggested that transport and diffusion of Cs would still benefit from the fluxing and volatile components.

7.5.4 Cesium Enrichment by Late Stage Hydrothermal Processes

Several observations within massive pollucite mineralisation allow the conclusion that other than magmatic processes can contribute to extreme enrichment of up to >30 wt% Cs_2O. Late stage hydrothermal processes are suggested to contribute to a major enrichment and redistribution of Cs enrichment at the Bikita and Tanco LCT pegmatite deposits (Teertstra and Černý 1995). According to Cruciani and Gualtieri (1999) tectosilicates of the analcime series are affected by a negative thermal expansion when cooled below 400 °C. The related volume loss is about 1 vol% and accompanied by the development of shrinking cracks. These cracks will open space for hydrothermal fluids and later will form the characteristic lepidolite vein network observed in massive pollucite mineralisation.The conditions of the thermal contraction at about 400 °C coincides well with the general transition from a magmatic to a hydrothermal dominated stage of the pegmatite formation (Sirbescu and

Nabelek 2003; London 2008). An opening of cracks within the massive pollucite mineralisation generates ideal pathways and a hydraulic sink for hydrothermal fluid.

With the late stage hydrothermal fluid, the Cs is transported along interstitials or fractures into the massive pollucite mineralisation where it is incorporated into the crystal lattice of Cs-analcime or Na-pollucite by $Cs + Al = Na + Si + OH$ coupled solid solution substitution. Experiments by Lagache (1995) and Lagache et al. (1995) could demonstrate that a continuous solid solution within the analcime-pollucite series is possible from about 330–600 °C. Other synthesis and substitution experiments (Barrer et al. 1953; Montagna et al. 2011) revealed that Cs incorporation into analcime is still effective at temperatures as low as 150 °C (Jing et al. 2017). Komarneni and Roy (1981) demonstrated that at these temperatures hydrothermal treatment of Cs chloride loaded zeolites resulted in the formation of pollucite. From the Yellowstone National Park, Keith et al. (1983) interpreted the 3000 ppm Cs within hydrothermal analcime to have been sourced by leaching of rhyolitic rocks with 2.5–7.6 ppm Cs. These hydrothermal temperature intervals are confirmed by the fluid inclusion study on pollucite from the Bikita LCT pegmatite deposit that reveal homogenisation temperatures ranging from 180 to 380 °C (Fig. 6.3). Although no experimental studies are performed until present, it is suggested that the substitution process further benefits from the effects of volatiles and fluxes, such as H_2O, B, F or P that increase the diffusion rates and solubility and make ion migration over greater distances possible (Simmons et al. 2003; Simmons 2007). When the remobilised Cs in the hydrothermal fluid meets no massive Na-pollucite zone, as at Catlin Creek and Wodgina, the Cs will be deposited along the rims of late stage minerals such as micas, tourmaline or beryl, or is disseminated along fractures or faults and/or dispersed to the surrounding host rocks.

The process of Cs leaching from feldspar rich lithologies and transport to the massive pollucite mineralisation requires the involvement of large amounts of hydrothermal fluids. Pegmatite fields like Bikita or Cattlin Creek pegmatite are emplaced along extensional tectonic structures (Fig. 2.2). These structures, in combination with the emplacement of an extreme volatile enriched pegmatite melt, can support the development of a local hydrothermal convection system. Deuteric fluids escaping from associated fertile granites most probably will use the same pathway through the fault system and will flow through the LCT pegmatite. These deuteric fluids may also carry incompatible elements like Cs and Rb towards the pegmatite system and may further enhance the hydrothermal replacement process.

On the other hand, escaping hydrothermal fluids from the pegmatite are able to change the composition of the host rock, and by circulation can get back into the pegmatite and modify the pegmatite composition. Galeschuk and Vanstone (2005) pointed out that the host rock lithology next to the Tanco LCT pegmatite deposit and related pegmatites is characterised by a prominent, up to 20 m thick aureole, that contains high concentrations of Li, Rb and Cs. Morgan and London (1987) could show that this metasomatic alteration of the amphibolite wall rock was even able to induce recrystallisation and the formation of the Li amphibole holmquistite. Also the interaction with host rock lithology favours the crystallisation of tourmaline within the border and wall zone of pegmatites (London 2008). Even though an interaction of host rock derived fluids with the pegmatite melt is possible, its effects on the composition of the pegmatite are considered to be low.

7.6 Genetic Model for the Formation of Massive Pollucite Mineralisation

The LCT pegmatites are considered to have been derived from highly fractionated and specialised granitic melts. Granitic melts can form either by anatexis in lower levels of a continental crust, or directly from melting of mantle rocks. Also, a direct generation of LCT pegmatitic melts by crustal anatexis without the intermediate stage of granitic melts could be possible (Fig. 7.5a). As large amounts of melt and hydrothermal fluids are required for the formation of LCT pegmatites with massive pollucite mineralisation, a genetic connection of the LCT pegmatite to a source granite is favoured here. The stable and radiogenic isotope data presented allow no safe discrimination of the source of the pegmatite melt that give rise to the formation of massive pollucite mineralisation. The field relationships clearly suggest an intrusive character of the LCT pegmatites. As LCT pegmatites form in late- to post-orogenic tectonic settings, emplacement primary takes place in extensional regimes with corresponding fractures and open spaces. Relationships in the field further suggest that the confining pressure of the host rocks must be high enough to withstand the internal pressure of the pegmatite melt. A large amount of melt is necessary that will intrude along in flat and gently dipping single structures in order to form a thick (>30 m) pegmatite sheet (Fig. 7.5b).

The initial crystallisation starts with the formation of the border and most probable the wall of the pegmatite (Fig. 7.5c). Emplacement of the melt into comparable cool host rocks will result in the formation of small scale hydrothermal convection cells that allow minor interaction of the pegmatite forming melt with the surrounding host rock. The crystallisation of the border and wall zones will further fractionate the remaining melt, with the incompatible elements (i.e., Cs, Rb) being enriched within the remaining melt. Due to ongoing fractionation, this melt may already contain up to 1–2 wt% Cs_2O.

After crystallisation of the border and wall zones the remaining melt will separate a relatively small portion (0.5–1 vol%) of immiscible melt and/or fluid that is extremely enriched in Al_2O_3 and Na_2O, as well as depleted in SiO_2 with a composition which allows the crystallisation of analcime (Fig. 7.5c). When the droplets of the segregating melt migrate, they absorb further Cs from the hosting melt. In the accumulated droplets, a concentration of 4–5 wt% Cs_2O appears feasible.

Fluxing components will undercool this Cs-analcime or Na-pollucite melt which prevents its crystallisation as intergranular grains. As this melt exhibits a lower relative density when compared to the remaining pegmatite melt, it will start to ascent gravitationally and accumulate within the upper portion of the pegmatite (Fig. 7.5d).

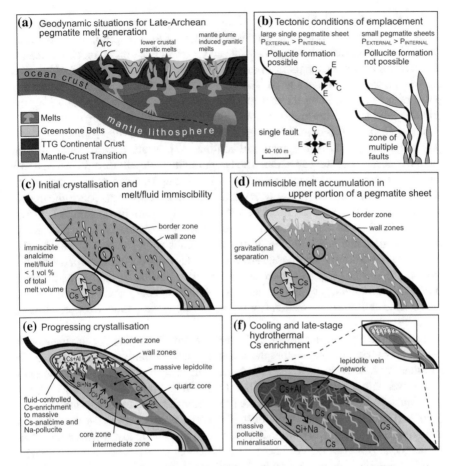

Fig. 7.5 Genetic model for the formation of massive pollucite mineralisations in LCT pegmatites. **a** Potential geotectonic scenarios for the formation of pegmatite melts. Archean continental crust with greenstone belts. Granitoid melt formation induced by underplating mantle lithosphere with oceanic crust. **b** Tectonic conditions of LCT pegmatite emplacement. Formation of massive pollucite mineralisations is favoured in flat lying and gently dipping large pegmatite sheets. **c** Initial crystallisation subsequent to magmatic melt fractionation and melt/fluid immiscibility with separation of a melt with analcime composition. Enrichment of Cs in immiscible analcime melt droplets. **d** Immiscible melt accumulation in upper portion of the pegmatite sheet, with further enrichment of Cs in melt droplets. **e** Final crystallisation of the pegmatite, with transition to a fluid-controlled Cs enrichment. Formation of cracks in massive Cs-analcime to Na-pollucite zone. **f** Cooling and late stage hydrothermal Cs-enrichment in massive pollucite mineralisation zone, with formation of lepidolite vein network

The reason for the initiation of this immiscibility process is unknown at present, but is suggested to be related to chemical changes of the composition of the initial melt. This compositional change is achieved by crystallisation of specific minerals that remove certain elements from the melt, or potentially also via the addition of elements from an external source. The latter is most probably related to additional pulse of new magma into the existing and still crystallising pegmatite, or by the addition of elements due to interaction with the host rocks of the pegmatites.

Final Crystallisation and Cooling: After exsolution and segregation of the Na-pollucite melt, the remaining melt will continue to crystallise separately and form the inner portions of the pegmatite. The Na-pollucite melt will then form a massive and almost monomineralic mineralisation within the upper portions of the pegmatite sheet (Fig. 7.5e). Contemporaneous crystallisation of the remaining pegmatite melt is still characterised by ongoing fractionation and enrichment of incompatible elements (e.g., Cs, Rb, Li) and concentration of these elements within exsolving hydrothermal fluids. As analcime and pollucite form a continuous solid solution series, the Na-pollucite melt is able to incorporate any available Cs from the melt and/or associated hydrothermal fluids and to crystallise in the upper portion of the pegmatite sheet (Fig. 7.5e). Continuing hydrothermal activity and ongoing substitution of Cs will then start to shift the composition from Cs-analcime or Na-pollucite towards a more pollucite compositions.

Late Stage Hydrothermal Cesium Enrichment: Even after the complete crystallisation of the LCT pegmatite, the remaining heat within the pegmatite and the surrounding host rocks will keep fluids flowing in hydrothermal circulation cells active over larger periods of time (10–30 Ma). When cooling below 400 °C, Cs-analcime and Na-pollucite are subjected to a negative thermal expansion of about 1 vol% (Cruciani and Gualtieri 1999). This volume loss causes the formation of shrinking cracks (Fig. 7.5f) within the future massive pollucite mineralisation. Consequently, the Cs bearing hydrothermal fluids can enter these voids and interact with the Na-pollucite. This interaction further enhances the coupled substitution and the gradual enrichment of Cs within the massive pollucite mineralisation. The opened cracks and circuits are later filled with late stage minerals such as lepidolite, quartz, feldspar and petalite and form the characteristic vein network.

During cooling of the pegmatite, late stage mineral replacement reactions (e.g., replacement of K-feldspar by lepidolite, cleavelandite and quartz), as well as sub-solidus self ordering processes in feldspars will release additional incompatible Cs and Rb into late stage hydrothermal fluids. As feldspar forms large portions of the pegmatite a considerable amount of Cs is released and circulates with the hydrothermal solutions towards the massive pollucite mineralisation zone (Fig. 7.5f).

At present it is not possible to ascertain the exact timing of the late stage Cs enrichment. Considering that within a substantial thick pegmatite sheet cooling proceeds steadily, starting from the border zone towards the inner portions of the pegmatite, it is assumed that the outer portions are already completely solidified, whereas the inner portions are still crystallising. Hence, the late stage replacement, subsolidus and hydrothermal processes are interpreted to already be active in the outer portion of the pegmatite sheet, whereas the inner portions are still crystallising.

When massive pollucite mineralisation bearing LCT pegmatites represent derivates from a source granite in close spatial relation, late hydrothermal fluids from the source granite will most probably ascent along the same structure in the crust that was used by the pegmatite melt. These fluids are then also interpreted to be able transporting incompatible elements like Cs and Rb, and will further contribute to the late-stage Cs enrichment in the massive pollucite mineralisation.

7.7 Implications to the Exploration for Cs-Bearing Pegmatites

Numerous Proterozoic, Paleozoic and Mesozoic LCT pegmatites are known, but these do not have the size and economic potential of the Archean ones, including massive pollucite mineralisation (cf. Dittrich et al. 2015). Thus the Meso- and Neo-Archean time window at 2850–2800 and 2650–2600 Ma, appears as most prospective, as geological conditions and factors that control the concentration of rare metal concentrations such as Li, Cs, Ta and Sn in pegmatites apparently have been realised there. All known large Archean LCT pegmatite provinces (Pilbara, Yilgarn, Zimbabwe, and Superior craton) are hosted by greenstone belt lithologies. LCT pegmatites are probably associated with granitoid suites with the same age span than LCT pegmatites, but until now no direct field evidence for a relationship to potential source granites could be observed.

Comparable geological settings and mineralogical and geochemical characteristics of other LCT economic pegmatite provinces worldwide (Bikita, Tanco) support the potential for the discovery of new deposits in the Pilbara and Yilgarn Craton that may contain economic quantities of massive pollucite mineralisation. Promising are flatly dipping single thick (>30 m) and zoned LCT pegmatite bodies.

In October 2016 the World's 3rd largest known pollucite deposit was discovered by Pioneer Resources Limited in the Yilgarn craton (WA) c. 140 km south of Kalgoorlie near Norseman during exploration for Li (spodumene). The discovery drilling PDRC015 shows 7 m @ 1.52 wt% Li_2O from 52 m and 6 m @ 27.7 wt% Cs_2O from 47 m (Pioneer Resources Limited 2016). In September 2018 Pioneer started to provide access to the pollucite ore body (10,500 t measured pollucite ore resource @ 17.1 wt% Cs_2O) located at depth of approximately 40 m below surface (Pioneer Resources Limited 2018). The Sinclair LCT pegmatite is hosted by mafic and ultramafic units of a greenstone belt and shows in the hanging wall a massive microcline zone and in the footwall a lepidolite zone (Crook 2018). The pollucite ore body is located in the centre of the Sinclair zoned LCT pegmatite and co-exists with petalite and lepidolite (Crook 2018).

References

Barrer RM, Baynham JW, McCallum N (1953) Hydrothermal chemistry of silicates. Part V. Compounds structurally related to analcite. J Chem Soc 832:4035–4041

Camacho A, Baadsgaard H, Davis D, Černý P (2012) Radiogenic isotope systematics of the Tanco and Silverleaf granitic pegmatites, Winnipeg River pegmatite district, Manitoba. Can Mineral 50:1775–1792

Černý P (1982) Petrogenesis of granitic pegmatites. In: Cerný P (ed) Short course in granitic pegmatites in science and industry, vol 8. Min Assoc Canada Short Course Handbook, pp 405–461

Černý P, Harris DC (1973) Allemontite and its alteration products from the Odd West pegmatite, southeastern Manitoba. Can Mineral 11:978–984

Černý P, Meintzer RE, Anderson AJ (1985) Extreme fractionation in rare element granitic pegmatites: selected examples of data and mechanisms. Can Mineral 23:381–421

Colombo F, Sfragulla J, Gonzáles del Tánago J (2012) The garnet-phosphate buffer in peraluminous granitic magmas: a case study from pegmatites in the Pocho district, Cordoba, Argentina. Can Mineral 50:1555–1571

Crook D (2018) Unearthing Australia's first pollucite deposit. RIU Explorer's Conference, Freemantle, Australia, 20–23 February 2018, extended abstract

Crouse RA, Černý P (1972) The Tanco pegmatite at Bernic Lake, Manitoba; I, Geology and paragenesis. Can Mineral 11:591–608

Cruciani G, Gualtieri A (1999) Dehydration dynamics of analcime by in situ synchrotron powder diffraction. Am Mineral 84:112–119

Dittrich T (2016) Meso- to Neoarchean Lithium-Cesium-Tantalum- (LCT-) pegmatites (Western Australia, Zimbabwe) and a genetic model for the formation of massive pollucite mineralisations. Dissertation Faculty of Geosciences, Geoengineering and Mining, TU Freiberg/Saxony, Germany, 341 pp. http://nbn-resolving.de/urn:nbn:de:bsz:105-qucosa-228968

Dittrich T, Seifert T (2013) Field work 2012—Cs-potential of LCT pegmatites in western Australia. Technical report, unpublished, prepared for: Rockwood Lithium GmbH, Frankfurt am Main. TU Bergakademie Freiberg, Division of Economic Geology and Petrology, 67 pp

Dittrich T, Seifert T, Schulz B (2015) Genesis of selected lithium-cesium-tantalum- (LCT) pegmatites of Western Australia—with special regards to their exploration potential for the Cs-mineral pollucite and additional data from field work in the Bikita LCT pegmatite field (Zimbabwe). Final technical report, unpublished, prepared for: Rockwood Lithium GmbH, Frankfurt am Main. TU Bergakademie Freiberg, Division of Economic Geology and Petrology, 536 pp and Appendix

Ellis AJ, Mahon WAJ (1977) Chemistry and hydrothermal systems. Academic Press, 392 pp

Endo A, Yoshikawa E, Muramatsu N, Takizawa N, Kawai T, Unuma H, Sasaki A, Masano A, Takeyamac Y, Kaharac T (2013) The removal of cesium ion with natural Itaya zeolite and the ion exchange characteristics. J Chem Technol Biotechnol 88:1597–1602

Ferguson LJ, Edgar AD (1978) The petrogenesis and origin of the analcime in the volcanic rocks of the Crowsnest I Formation, Alberta. Can J Earth Sci 15:69–77

Ferreira KJ (1984) The mineralogy and geochemistry of the lower Tanco Pegmatite, Bernic Lake, Man., Canada. M.Sc. thesis, University of Manitoba Winnipeg, Canada, 256 pp (unpublished)

Fowler AD, Doig R (1983) The significance of europium anomalies in the REE spectra of granites and pegmatites, Mont Laurier, Quebec. Geochim Cosmochim Acta 47:1131–1137

Galeschuk CR, Vanstone PJ (2005) Exploration for buried rare-element pegmatites in the Bernic Lake area of southern Manitoba. In: Linnen RL, Samson IM (eds) Rare-Element Geochemistry and Mineral Deposits. Geol Assoc Canada Short Course Notes 17:159–173

Goscombe B, Blewett RS, Czarnota K, Groenewald PB, Maas R (2009) Metamorphic evolution and integrated terrane analysis of the eastern Yilgarn Craton: rationale, methods, outcomes and interpretation. Geol Surv West Austral Record 23, 281 pp

Grosse P, Toselli AJ, Rossi JN (2010) Petrology and geochemistry of the orbicular granitoid of Sierra de Velasco (NW Argentina) and implications for the origin of orbicular rocks. Geol Mag 147:451–468

Gwavava O, Ranganai RT (2009) The geology and structure of the Masvingo greenstone belt and adjacent granite plutons from geophysical data, Zimbabwe Craton. South Afr J Geol 112:277–290

Inagaki Y, Shinkai A, Idemistu K, Arima T, Yoshikawa H, Yui M (2006) Aqueous alteration of Japanese simulated waste glass P0798: effects of alteration-phase formation on alteration rate and cesium retention. J Nucl Mater 354:171–184

Jing Z, Cai K, Li Y, Fan J, Zhang Y, Miao J, Chen Y, Jin F (2017) Hydrothermal synthesis of pollucite, analcime and their solid solutions and analysis of their properties. J Nucl Mater 488:63–69

Keith TEC, Thompson JM, Mazs RE (1983) Selective concentration of cesium in analcime during hydrothermal alteration, Yellowstone National Park, Wyoming. Geochim Cosmochim Acta 47:795–804

Komarneni S, Roy R (1981) Zeolithes for fixation of Cesium and Strontium from radwastes by thermal and hydrothermal treatments. Nucl Chem Waste Man 2:259–264

Lagache M (1995) New experimental data on the stability of the pollucite-analcime series: application to natural assemblages. Eur J Mineral 7:319–324

Lagache M, Dujon SC, Sebastian A (1995) Assemblages of Li-Cs pegmatite minerals in equilibrium with a fluid from their primary crystallization until their hydrothermal alteration: an experimental study. Mineral Petrol 55:131–143

Liou JG (1971) Analcime equilibria. Lithos 4:389–402

London D (1984) Experimental phase equilibria in the system $LiAlSiO_4$–SiO_2–H_2O: a petrogenetic grid for lithium-rich pegmatites. Am Mineral 69:995–1004

London D (1987) Internal differentiation of rare element pegmatites: effects of boron, phosphorus and fluorine. Geochim Cosmochim Acta 51:403–420

London D (2008) Pegmatites, vol 10. Spec Publ Can Mineral, 368 pp. ISBN 978-0-921294-47-4

London D, Morgan GBVI, Babb HA, Loomis JL (1993) Behavior and effects of phosphorus in the system Na_2O–K_2O–Al_2O_3–SiO_2–P_2O_5–H_2O at 200 MPa(H_2O). Contrib Mineral Petrol 113:450–465

London D, Morgan GBV, Icenhower J (1998) Stability and solubility of pollucite in the granite system at 200 MPa H_2O. Can Mineral 36:497–510

London D, Morgan GBV, Icenhower J (2017) Erratum: stability and solubility of pollucite in the granite system at 200 MPa H_2O. Can Mineral 55:945–946

London D, Wolf MB, Morgan GB, Garrido MG (1999) Experimental Silicate-Phosphate equilibria in Peraluminous Granitic Magmas, with a case study of the Alburquerque Batholith at Tres Arroyos, Badajoz. Spain J Petrol 40(1):215–240

Lux DR, Hooks B, Gibson D, Hogan JP (2007) Magma interactions in the Deer Isle granite complex, Maine; field and textural evidence. Can Mineral 45:131–146

Martin H, Smithies RH, Rapp R, Moyend JF, Champion D (2005) An overview of adakite, tonalite-trondhjemite-granodiorite (TTG), and sanukitoid: relationships and some implications for crustal evolution. Lithos 79:1–24

McLennan SM, Hemming S, McDaniel DK, Hanson GN (1993) Geochemical approaches to sedimentation, provenance, and tectonics. Geol Soc Am Spec Pap 284:21–40

Melcher F, Graupner T, Gäbler HE, Sitnikova M, Henjes-Kunst F, Oberthür T, Gerdes A, Badanina E, Chudy T (2017) Mineralogical and chemical evolution of tantalum-(niobium-tin) mineralisation in pegmatites and granites. Part 2: Worldwide examples (excluding Africa) and an overview of global metallogenetic patterns. Ore Geol Rev 89:946–987. https://doi.org/10.1016/j.oregeorev.2016.03.014

Montagna G, Arletti R, Vezzalini G, di Renzo F (2011) Borosilicate and aluminosilicate pollucite nanocrystals for the storage of radionuclides. Powder Technol 208:491–495

Morgan GB, London D (1987) Alteration of amphibolitic wallrocks around the Tanco rare-element pegmatite, Bernic Lake, Manitoba. Am Mineral 72:1097–1121

Mueller AG, Hagemann SG, McNaughton NJ (2016) Neoarchean orogenic, magmatic and hydrothermal events in the Kalgoorlie-Kambalda area, Western Australia: constraints on gold mineralization in the Boulder Lefroy-Golden Mile fault system. Miner Deposita 51:1–31. https://doi.org/10.1007/s00126-016-0665-9

Pearce TH (1970) The analcite-bearing volcanic rocks of the Crowsnest Formation, Alberta. Can J Earth Sci 7:46–66

Peters TJ, Luth WC, Tuttle OF (1966) The melting of analcite solid solutions in the system $NaAlSiO_4$–$NaAlSi_3O_8$–H_2O. Am Mineral 51:736–753

Pioneer Resources Limited (2016) Drilling results, ASX announcement, 4 October 2016 http://www.pioneerresources.com.au/downloads/asx/pio2016100401.pdf

Pioneer Resources Limited (2018) Commences mining operation at Sinclair caesium mine, 13 September 2018 http://www.pioneerresources.com.au/downloads/asx/pio2018121201.pdf

Pistone M, Arzilli F, Dobson KJ, Cordonnier B, Reusser E, Ulmer P. Marone F, Whittington AG, Mancini L, Fife JL, Blundy JD (2015) Gas-driven filter pressing in magmas: Insights into in-situ melt segregation from crystal mushes. Geology 43:699–702

Propach G (1976) Models of filter differentiation. Lithos 9:203–209

Redkin HF, Hemley JJ (2000) Experimental Cs and Sr sorption on analcime in rockbuffered systems at 250–300 °C and Psat and the thermodynamic evaluation of mineral solubilities and phase relations. Eur J Mineral 12:999–1014

René M (2012) Occurence of Th, U, Y, Zr, and REE-bearing acessory minerals in granites and their petrogenetic significance. In: Blasik M, Hanika B (eds) Granite—occurence, mineralogy and origin. Earth Science in the 21st Century, Nova Sciences Publishers, New York, pp 27–56

Ribbe PH (1983) Aluminium-silicon order in feldspars: domain textures and diffraction patterns. In: Ribbe PH (ed) Feldspar mineralogy. Rev Mineral 2:21–54

Richter L, Seifert T, Dittrich T, Schulz B, Hagemann S, Banks D (2015a) Constraints on the magmatic-hydrothermal fluid evolution in LCT pegmatites from Mt. Tinstone, Wodgina Pegmatite District, North Pilbara Craton, Western Australia. Mineral resources in a sustainable world, 13th SGA Biennial Meeting 2015 Nancy, Proceedings vol 2, pp 529–532

Richter L, Lüders V, Hagemann SG, Seifert T, Dittrich T (2015b) Stable carbon isotopic composition of fluid inclusions from the Archean Bikita LCT pegmatite field. GeoBerlin 2015-Dynamic Earth from Alfred Wegener to today and beyond, 4–7 October 2015, GFZ German Research Centre for Geosciences, Berlin. GFZ Abstracts, pp 310–311. https://doi.org/10.2312/gfz.lis.2015.003

Roux J, Hamilton L (1976) Primary igneous analcite—an experimental study. J Petrol 17:244–257

Sanchez-Munoz L, Gracia-Guinea J, Zagorsky VY, Juwono T, Modereski PJ, Cremades A, Van Tendeloo G, de Moura OJM (2012) The evolution of twin patterns in perthitic K-feldspar from granitic pegmatites. Can Mineral 50:989–1024

Sebastian A, Lagache M (1990) Experimental study of the equilibrium between pollucite, albite and hydrothermal fluid in pegmatitic systems. Mineral Mag 54:447–454

Seifert T (2008) Metallogeny and petrogenesis of lamprophyres in the Mid-European Variscides—post-collisional magmatism and its relationship to late-Variscan ore forming processes (Bohemian Massif). IOS Press BV, Amsterdam, p 303

Selway JB, Breaks FW, Tindle AG (2005) A review of rare-element (Li–Cs–Ta) pegmatite exploration techniques for the Superior Province, Canada, and large worldwide tantalum deposits. Explor Min Geol 114:1–30

Simmons WB (2007) Gem bearing pegmatites. In: Groat LA (ed) Geology of gem deposits, vol 37. Mineral Assoc Canada Short Course Series, pp 169–206

Simmons WB, Webber KL, Falster AU, Nizamoff JW (2003) Pegmatology—pegmatite mineralogy, petrology and petrogenesis. Rubellite Press, New Orleans, 176 pp

Simpson FM (1974) The mineralogy of pollucite and beryl from the Tanco Pegmatite at Bernic Lake, Manitoba. M.Sc.-Thesis University of Manitoba, Winnipeg, Canada, 105 pp (unpublished)

Sirbescu MLC, Nabelek PI (2003) Crustal melts below 400 °C. Geology 31:685–688

Sisson TW, Bacon CR (1999) Gasdriven filter pressing in magmas. Geology 27:613–616

Stilling A, Černý P, Vanstone PJ (2006) The Tanco pegmatite at Bernic Lake, Manitoba; XVI, Zonal and bulk compositions and their petrogenetic significance. Can Mineral 44:599–623

Sweetapple MT, Collins PLF (2002) Genetic framework for the classification and distribution of Archean rare metal pegmatites in the North Pilbara Craton, Western Australia. Econ Geol 97:873–895

Taylor SR, McLennan SM (1985) The continental crust; its composition and evolution; an examination of the geochemical record preserved in sedimentary rocks. Blackwell, Oxford, p 312

Teertstra DK, Černý P (1995) First natural occurrence of end-member pollucite: a product of low-temperature reequilibration. Eur J Mineral 7:1137–1148

Thomas R, Davidson P (2012) Water in granite and pegmatite-forming melts. Ore Geol Rev 46:32–46

Thomas R, Webster JD, Davidson P (2011) Be-daughter minerals in fluid and melt inclusions: implications for the enrichment of Be in granite-pegmatite systems. Contrib Mineral Petrol 161:483–495

Tkachev AV (2011) Evolution of metallogeny of granitic pegmatites associated with orogens through geological time. In: Sial AN, Bettencourt JS, de Campos CP, Ferreira VP (eds) Granite-related ore deposits. Geol Soc London Spec Publ 350:7–23

Trueman DL, Černý P (1982) Exploration for rare-element granitic pegmatites. In: Cerný P (ed) Short course in granitic pegmatites in science and industry, vol 8. Min Assoc Canada Short Course Handbook, pp 463–494

Vasyukova O, Williams-Jones AE (2014) Fluoride-silicate melt immiscibility and its role in REE ore formation: evidence from the Strange Lake rare metal deposit, Québec-Labrador, Canada. Geochim Cosmochim Acta 139:110–130

Woolley AR, Symes RF (1976) The analcime-phyric phonolites (blairmorites) and associated analcime kenytes of the Lupata Gorge, Mocambique. Lithos 9:9–15

Yokomori Y, Asazuki K, Kamiya N, Yano Y, Akamatsu K, Toda T, Aruga A, Kaneo Y, Matsuoka S, Nishi K, Matsumoto S (2014) Final storage of radioactive cesium by pollucite hydrothermal synthesis. Scientific Reports 4-4195, 4 pp. https://doi.org/10.1038/srep04195

Yund RA (1983) Microstructures, kinetics and mechanism of alkali feldspar exsolution. In: Ribbe PH (Ed) Feldspar Mineralogy. Rev Mineral 2:177–222

Printed in the United States
By Bookmasters